Introduction to

Bioinstrumentation

CONTEMPORARY INSTRUMENTATION AND ANALYSIS

Edited by: Gary M. Hieftje

Introduction to
Bioinstrumentation

With Biological, Environmental, and Medical Applications

Clifford D. Ferris
Bioengineering Program
University of Wyoming
Laramie, Wyoming

THE HUMANA PRESS • CLIFTON, NEW JERSEY

Library of Congress Catalog Card No.: 78–0579
Ferris, Clifford D.
    Introduction to Bioinstrumentation.
Clifton, N.J.: Humana Press
352 p.
7811   780912

We gratefully acknowledge permission from the Plenum Press to draw extensively
in Chapters 6 and 7 from the author's earlier work: Introduction to Bioelectrodes,
Plenum Press, New York, 1974.

ISBN: 0-89603-000-8   Regular Edition
ISBN: 0-89603-000-2   Soft Cover Text Edition

Printed in the United States of America

Preface

The present volume is designed as a practical tutorial survey not only for all those interested in bioinstrumentation and its applications, but also as a text for a one-semester upper-division undergraduate course in instrumentation for bioengineering students. A knowledge of basic physics, basic electronics, and mathematics to elementary linear differential equations is assumed. The book is well suited for use as a reference source for all research and clinical workers in the fields of biology, medicine, and the environmental sciences who have an adequate background in the physical sciences. At the University of Wyoming, the text is also used for a course in the interdisciplinary program for graduate study in the neurosciences.

The philosophy espoused herein is fundamental system analysis and design, rather than detailed discussion of particular devices produced by commercial manufacturers. Equipment-oriented texts, although initially useful, tend to become obsolete rather rapidly. Basic design and analysis techniques change little with time. Discussion has been limited to devices that have found applications in the biological, environmental, and medical fields. Many transducers used in other disciplines have been omitted. It is not the author's intent to produce a compendium of transducer applications, but rather an introduction to those techniques used in the environmental, biological, and medical sciences.

Only analog systems are described, although occasional references are made to digital systems and techniques. The philosophy underlying this choice is quite simple. Nature and biological systems are inherently analog in behavior, although arguments can be advanced for some form of digital processing in the central nervous system. Bioelectric potentials, biological growth and decay, and environmental processes represent analog system functions. Computer applications in medicine and

v

biology have been adequately described by other authors, and for this reason, various topics have been omitted from this book. These include a number of ultrasonic diagnostic techniques and X-ray tomography, among others. Digital circuit design is more than adequately covered in existing texts.

With respect to subject matter, the book is divided into six Sections. Section 1 examines a general second-order linear system with regard to overall system response. The concepts of natural and damped frequencies of oscillation, damping, risetime, and time constants are introduced. Properties of nonlinear systems such as hysteresis, saturation, and deadband are discussed as well.

The following three Sections are arranged in the same sequence as the components in an instrumentation system. Section 2 explores transducers with respect to method of operation and application to biomedical and environmental problems. Section 3 examines the requirements for amplifiers and signal processing equipment relative to the transducer systems discussed in the previous section. Section 4 treats recording and display equipment, including oscillographs, chart recorders and FM tape systems.

The basic design and performance criteria for biotelemetry systems are presented in Section 5. In Section 6 and the final chapter we discuss some practical matters involving electrical hazards, grounding, shielding and equipment selection. A final Appendix presents a rigorous mathematical approach to overall analysis of a hypothetical instrumentation system from a general system theory point-of-view. Also included is a short set of exercise problems. It is difficult to develop a problem set for a text of this nature. The problems included reflect the basic subjects treated in the text and should form a base from which course instructors can develop additional problems to suit their particular requirements.

Fom time to time, reference is made to various commercially available devices. Such comments only indicate the types of devices available and do not constitute an endorsement of specific products, nor lack of endorsement of others.

September 1978   Clifford D. Ferris
                                                 Laramie, Wyoming

Acknowledgements

Chapters 6 and 7 draw heavily upon material presented by the author in an earlier work: *Introduction to Bioelectrodes*, Plenum Press, New York, 1974. The material in Chapter 6 has been excerpted from the overall coverage in this prior work. Chapter 7 duplicates much of the same chapter in that book, although some subjects have been deleted and much new material has been added.

The author would like to acknowledge various contributions to this work. Ms. Georgia Prince of Plenum Publishing Corporation kindly granted permission for the use of the material from the author's previous book: *Introduction to Bioelectrodes*. Several industrial organizations supplied photographs and information concerning their products and are so recognized in the appropriate figure captions. The development of the electronic instruments used for avalanche prediction (Chapter 10) was in part supported by grants to the University of Wyoming from the U.S. Forest Service and the National Park Service. The scanning densitometer system described in Chapter 7 was constructed by Alonna S. Widdoss as part of a project toward her Master of Science in Bioengineering degree at the University of Wyoming. Jacqueline I. Howell programmed the computer plot routines for the system response curves contained in the Appendix. The author's students in BE 691D at the University of Wyoming for several years served as guinea pigs regarding preliminary versions of the manuscript. Many useful suggestions were made by them.

Drs. James E. Lindsay, C. Norman Rhodine, and Professor Richard W. Weeks kindly read selected portions of the text and made helpful suggestions. Sue Bruhnke and Shelley Bryant expertly typed the manuscript and many thanks are due them. Special thanks are owed Thomas Lanigan of The Humana Press for his continued encouragement during the manuscript preparation.

Technical Note

Throughout this text, the systems of units are those presently appropriate for the devices discussed. Where applicable, SI units have been adopted. Despite Congressional action, however, the metric system is being adopted very slowly in the United States and there is mounting opposition to its use from various groups. In medicine, for example, blood pressure is traditionally expressed in millimeters of mercury (mm Hg) and not in the SI unit of pascals. For this reason, the SI system purposely has not been used uniformly in this book

In many of the illustrations, electronic amplifiers are shown. Generally they appear in functional form with respect to signal paths. It must be remembered that DC bias potentials must also be supplied in order to achieve correct device operation. For the most part, power supply (bias) and AC power line connections have been omitted. This is standard practice in books on electronics where signal analysis is the primary concern.

Contents

Section 3:   Amplifiers and Signal Conditioning

Chapter 7.   Preamplifiers for Use with Transducers and Bioelectrodes   **151**

Section 6:   Practical Matters

Section 1

GENERAL SYSTEMS

1

Behavior of Linear Systems

1.1. INTRODUCTION

In the physical world, we can identify two mechanisms by which energy is stored and one linear process by which it is dissipated. Potential energy is stored, for example, by virtue of the position of a mass above the earth's surface, the extension of a spring, the height of a fluid column, or the electric charge stored in a capacitor. Kinetic energy is stored by virtue of a moving mass, such as a flywheel, the winding of a spring, fluid flow, or the magnetic field associated with a steady electrical current in an inductance coil. Energy dissipation results in the production of electromagnetic radiation, usually in the form of heat. The principal dissipative mechanism is friction, either viscous or dry. The latter is a nonlinear process, whereas the former is linear.

FIG. 1.1   The basic linear system.

A physical system can be represented by the block diagram ("black box") shown in Fig. 1.1, where $e(t)$ is the time excitation to the system and $r(t)$ is the time response of the system to that excitation. If an excitation $e_i(t)$ produces a response $r_i(t)$, then two necessary conditions for a linear system are:

$$\sum_i e_i(t) \to \sum_i r_i(t) \tag{1.1}$$

$$k \sum_i e_i(t) \to k \sum_i r_i(t) \tag{1.2}$$

Equation (1.1) is known as *the principle of superposition*, while Eq. (1.2) is called *the preservation of scale factor*. The quantity k is an arbitrary constant. The necessary and sufficient conditions for a linear system are: (1) validity of the principle of superposition and (2) preservation of scale factor. Linear systems are described mathematically by linear algebraic equations, linear difference equations, or linear differential equations.

Viscous friction is described mathematically by the relationship

$$\mathbf{F} = R_f \mathbf{u} \tag{1.3}$$

in which $\mathbf{F}$ is restoring or applied force (a vector), $\mathbf{u} =$ applied or resultant velocity (a vector), and R_f is the coefficient of viscous friction. Figure 1.2 illustrates graphically the relation expressed in Eq. (1.3). Dry friction, on the other hand, is a nonlinear process, as shown in Fig. 1.3. Symbolically, it is defined by

$\mu_0 =$ coefficient of static friction (case of no relative motion of the associated physical bodies)

$\mu_k =$ coefficient of kinetic (sliding) friction (case when relative motion occurs between the associated bodies)

No simple relationship of the kind shown in Eq. (1.3) can be written for the dry friction case.

A physical law states that energy must be conserved. Thus for a physical system

applied energy = energy stored + energy dissipated

Kirchhoff's Laws, derived from the principle of conservation of energy, and d'Alembert's principle, state that in a closed system

$$\sum_k f_k = 0 \qquad \sum_k v_k = 0 \tag{1.4}$$

where f_k represents all of the forces (applied and restorative) associated

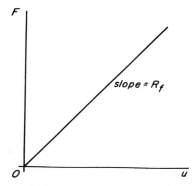

FIG. 1.2   Force–velocity relationship for viscous (linear) friction.

with a mechanical system; v_k represents all of the electrical voltage drops (with appropriate algebraic sign) associated with a closed loop in an electrical system.

Thus when we write the force balance for a mechanical system or the voltage-drop balance for a closed loop in an electrical system as expressed in Eq. (1.4), we have

$$f(t) = M\frac{du}{dt} + R_m u + K \int u\, dt$$

$$v(t) = L\frac{di}{dt} + Ri + \frac{1}{C} \int i\, dt$$

(1.5)

where u and i (respectively, mechanical velocity and electric current) are functions of time, and M, R_m, K, L, R, and C are time invariant. In

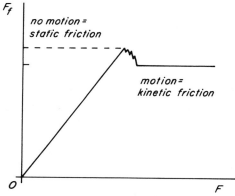

FIG. 1.3   Force relationship for static (dry) friction: F = applied force; F_f = dry friction force.

formulating Eq. (1.5), a unidirectional force–velocity relationship has been assumed for which the constituent relations are:

$$f = M \frac{du}{dt} \qquad \text{Newton's 2nd Law}$$

$$f = R_m u$$

$$f = Kx = K \int u \, dt \qquad \text{Hooke's law } (x = \text{displacement})$$

$$v = L \, di/dt \qquad \text{derived from Faraday's Law}$$

$$v = Ri \qquad \text{Ohm's Law}$$

$$v = 1/C \int i \, dt \qquad \text{derived from Coulomb's Law}$$

If it is assumed that $f(t)$ and $v(t)$ are mathematically differentiable, the Eqs. (1.5) can be expressed in the generalized form

$$\frac{d^2 r}{dt^2} + 2\zeta\omega_0 \frac{dr}{dt} + \omega_0{}^2 r = \epsilon(t) \tag{1.6}$$

where $\epsilon(t)$ is the mathematical excitation function [the first derivative of the physical excitation function $e(t)$] and $r = r(t)$ is the response to $e(t)$. The system damping factor is ζ, and the natural frequency of the system as derived subsequently is ω_0. In terms of the physical systems expressed by the relations in Eqs. (1.5), ζ and ω_0 have the following values: $\zeta = R_m/2M\omega_0 = R/2L\omega_0$, and $\omega_0 = \sqrt{K/M} = 1/\sqrt{LC}$.

Since Eq. (1.6) is second order, there are three possible solutions. These solutions depend upon the values of the physical constants contained within ζ and ω_0. The general solution to the homogeneous form of Eq. (1.6), i.e., $\epsilon(t) \equiv 0$ is given by

$$r_c(t) = k_1 e^{d_1 t} + k_2 e^{d_2 t} \tag{1.7}$$

The homogeneous response $r_c(t)$, also known in mathematics as the complementary function, represents the natural or transient response of the system, that is the response to shock or impulse excitation. An impulse is defined by the Dirac δ-function

$$\int_{-\infty}^{\infty} \delta(t) \, dt = 1 \tag{1.8}$$

After the initial impulse, the system is at liberty to respond in its natural mode relative to the parameters ζ and ω_0.

1.2. NATURAL SYSTEM RESPONSE OF THE SECOND-ORDER LINEAR SYSTEM

Equation (1.7) represents the natural or free response of a second-order linear system to a shock excitation, $\epsilon(t) = \delta(t)$, which corresponds in classical mathematics to $\epsilon(t) = 0$. At this point k_1 and k_2 are constants of integration that must be evaluated later based upon the system boundary conditions. The roots d_1 and d_2 have the form

$$d_1 = -\zeta\omega_0 + \omega_0\sqrt{\zeta^2 - 1} \qquad d_2 = -\zeta\omega_0 - \omega_0\sqrt{\zeta^2 - 1} \quad (1.9)$$

In order to develop an understanding of the natural behavior of a linear second-order system, we now examine the behavior of the roots d_1, d_2 as a function of the system damping factor ζ, as ζ varies from 0 (no damping) to ∞ (total damping). To simplify the ensuing discussion we write the roots in terms of a general complex number having a real part σ and an imaginary part $j\omega$. Writing the roots in this form facilitates a complex plane plot of the locus of the roots. With this new notation, the roots have the form s_1, $s_2 = \sigma \pm j\omega$, or s_1, $s_1^* = \sigma \pm j\omega$, where (*) indicates a complex conjugate. When $\zeta = 0$, then s_1, $s_1^* = \pm j\omega_0$. The roots are purely imaginary.

In the range $0 < \zeta < 1$,

$$s_1, s_1^* = -\omega_0\zeta \pm j\omega_0\sqrt{1 - \zeta^2} = \sigma \pm j\omega$$

In this range of ζ, a conjugate complex pair of roots occurs. Note, however, that the magnitude of the roots ($|s_1|$, $|s_1^*|$) is ω_0.

When $\zeta \equiv 1$, then s_1, $s_2 = -\omega_0$ producing real and equal roots.

For the final case when $1 < \zeta < \infty$,

$$s_1, s_2 = -\omega_0\zeta \pm \omega_0\sqrt{\zeta^2 - 1}$$

the roots are real and unequal. The root s_1 increases in the positive direction to an upper bound of 0 as $\zeta \to \infty$. The root s_2 increases without bound in negative value as $\zeta \to \infty$.

The behavior of the roots s_1, s_2 is portrayed graphically in Fig. 1.4 as a *root-locus* plot. The plane employed for the plot is the *s*-plane, consisting of an axis of reals, σ, and an axis of imaginaries, $j\omega$.

Root-locus plots of the sort shown in Fig. 1.4 are extremely useful in determining the stability of a system, and find extensive use in the study of control systems. The plots are not restricted to second-order

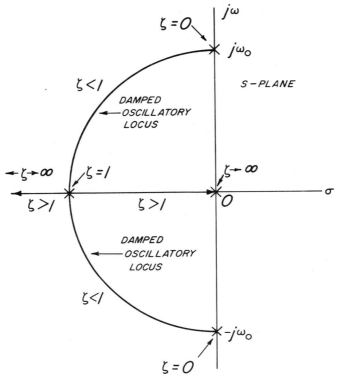

FIG. 1.4   Root-locus plot of the roots of the linear second-order system equation.

systems. In terms of system response, the root-locus plot shown above is interpreted in the following manner: When $\zeta = 0$, a free oscillatory system response occurs; there is no damping. As ζ increases, the roots move from $+j\omega_0$ and $-j\omega_0$ along a circular path until they meet on the negative real axis. At the critical point $\sigma = -\omega_0$ the response ceases to be damped oscillatory and becomes critically damped. As ζ increases further, one root moves to the origin, while the other tends to infinity along the negative real axis. This last behavior accounts for the over-damped response. The associated behaviors of $r(t)$ as a function of time are illustrated in Fig. 1.5.

The quantity $\omega_0 \sqrt{1 - \zeta^2}$ for $0 < \zeta < 1$ is sometimes referred to as the damped frequency of response ω_d, ($\omega_d < \omega_0$), while ω_0 is the un-damped or free response; ω_d decreases as ζ increases, reaching a limiting value of 0 when ζ increases to the critical value of unity. Figure 1.6 illustrates a simple geometric relationship between ω_0 and ω_d.

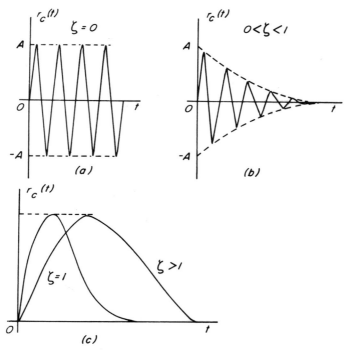

FIG. 1.5 Response with ζ as a parameter. In (a), $r_c(t) = A \sin \omega_0 t$; in (b), $r_c(t) = Ae^{-\zeta \omega_0 t} \sin \omega_0 \xi t$, where $\xi = \sqrt{1 - \zeta^2}$; in (c), $r_c(t) = (A + Bt)e^{-\omega_0 t}$, ($\zeta = 1$); $r_c(t) = A_1 e^{-(\omega_0 \zeta - \xi)t} - A_2 e^{-(\omega_0 \zeta + \xi)t}$, ($\zeta > 1$).

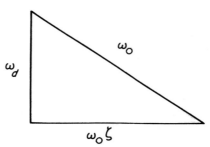

FIG. 1.6 Right-triangle frequency relation: $\omega_d = \xi \omega_0 = \omega_0 \sqrt{1 - \zeta^2}$; $\omega_d^2 = \omega_0^2 - \omega_0^2 \zeta^2$; from the Pythagorean theorem $\omega_0^2 = \omega_d^2 + \omega_0^2 \zeta^2$.

1.3. FORCED EXCITATION IN LINEAR SECOND-ORDER SYSTEMS

In most second-order system analysis, the excitations $e(t)$ and $\epsilon(t)$ are not zero, but some prescribed function of time. The complete system response is then composed of the complementary function and the particular integral, that is,

$$r(t) = r_c(t) + r_p(t)$$

where $r_p(t)$ is the response of the system to some driving (excitation) function that is nonzero. It is not possible to develop a general particular integral solution as we did for the complementary function in Eq. (1.6). In physical terms, $r_p(t)$ represents the forced or driven response of the system, while $r_c(t)$ is the natural or free response.

To illustrate the method of solution, we now resort to an example. Let us assume some second-order system to be defined by

$$\frac{d^2r}{dt^2} + 3\frac{dr}{dt} + 2r = K_0$$

and subject to the boundary conditions $r = 0$ when $t = 0$, and $r = K_0/2$ when $t = 1$. For this case, $2\zeta\omega_0 = 3$ and $\omega_0{}^2 = 2$; thus $\zeta = 1.06$. Since $\zeta > 1$, the natural response will be overdamped, and the roots d_1, d_2 are found using operational notation in which $d/dt = D$. The homogeneous portion of the system equation is rewritten as

$$[D^2 + 3D + 2]r_c = 0$$

For a nontrivial solution, $r_c \neq 0$, therefore

$$D^2 + 3D + 2 = 0.$$

Solving this quadratic equation in D for its roots yields $d_1 = -1$ and $d_2 = -2$. From Eq. (1.7),

$$r_c(t) = k_1 e^{-t} + k_2 e^{-2t}$$

The forced response is found using the *method of undetermined coefficients* in which we assume $r_p(t)$ to have the same form as the driving function K_0. Let $r_p(t) = a_0$. Thus

$$\frac{d^2 a_0}{dt^2} + 3\frac{da_0}{dt} + 2a_0 = K_0$$

But since a_0 equals a constant, $2a_0 = K_0$, and $a_0 = K_0/2$. Combining the free and forced responses yields

$$r(t) = r_c(t) + r_p(t) = k_1 e^{-t} + k_2 e^{-2t} + K_0/2$$

To evaluate the integration constants k_1 and k_2, we substitute the boundary conditions, independently, into the response function $r(t)$. This yields two algebraic equations which are solved simultaneously for k_1 and k_2.

$$k_1 + k_2 = -K_0/2 \qquad (t = 0)$$
$$0.368k_1 + 0.135k_2 = 0 \qquad (t = 1)$$
$$k_1 = 0.29K_0 \qquad k_2 = -0.79K_0$$

Substitution of these values of k_1 and k_2 into the solution $r(t)$ yields the complete response

$$r(t) = K_0(0.5 + 0.29e^{-t} - 0.79e^{-2t})$$

The system response consists of a constant value, modified by an initial overdamped transient, as shown in Fig. 1.7.

The more usual system theory approach is to use the method of Laplace transforms to find the solution to differential equations. When this method is used, $r_c(t)$ and $r_p(t)$ are found simultaneously. For the system under discussion, the Laplace transform solution would proceed as follows:

$$s^2 \mathcal{R}(s) + 3s\mathcal{R}(s) + 2\mathcal{R}(s) = \frac{K_0}{s} + r'(0) + sr(0) + 3r(0)$$

$$\mathcal{R}(s) = \frac{K_0 + sr'(0)}{s(s^2 + 3s + 2)} \qquad \text{[in which } \mathcal{L}[r(t)] = \mathcal{R}(s)\text{]}$$

$$= \frac{K_0}{s(s + 1)(s + 2)} + \frac{r'(0)}{(s + 1)(s + 2)}$$

[in which $r(0) = 0$ from boundary conditions]

From inverse transform tables we can evaluate $r(t)$.

$$r(t) = K_0\left(\frac{1}{2} - e^{-t} + \frac{e^{-2t}}{2}\right) + r'(0)(e^{-t} - e^{-2t})$$

The value of $r'(0)$ is evaluated from the boundary condition, $r(1) = K_0/2$, which yields $r'(0) = 1.29K_0$. Thus

$$r(t) = K_0(0.5 + 0.29e^{-t} - 0.79e^{-2t})$$

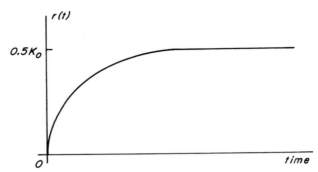

FIG. 1.7   Response of example system to applied constant excitation.

1.4.   TIME CONSTANTS

An important factor in system theory is the time required for the system to respond to a given excitation. The response damping factor (as opposed to the system damping factor ζ) is given by

$$e^{-\omega_0 \zeta t}$$

and it directly affects the time required for the transient to decay to zero. The product $\omega_0 \zeta$ is the decrement factor or attenuation factor, and has the dimensions of reciprocal time. The inverse of the decrement factor is called the time constant, and is a measure of the time required for the transient response to decay to $1/e$ of its initial value. If we take the first derivative of the function

$$f(t) = e^{-\omega_0 \zeta t}$$

we can find the rate of change of $f(t)$. Thus

$$f'(t) = -\omega_0 \zeta e^{-\omega_0 \zeta t} = -\omega_0 \zeta f(t)$$

The slope of $f(t)$ is the same as its rate of change and has the value $-\omega_0 \zeta$. The rate of change is negative, signifying decay of the function $f(t)$. The initial rate of change is given by the derivative $f'(t)$ evaluated at time $t = 0$. The initial rate of change is $-\omega_0 \zeta$, and would be the slope of the decay curve if the decay were a linear function of time. The reciprocal of this initial slope represents the time that would be required for the function to decay to zero if the decay continued at the initial rate. Exponential functions do not decay linearly and a function of the form $\exp(-t/\tau)$ decays in time $t = \tau$ seconds to $1/e$ or 0.3679 of its initial value at time $t = 0$. The quantity $\tau = 1/\omega_0 \zeta$ is the time constant. Figure 1.8 illustrates the present discussion.

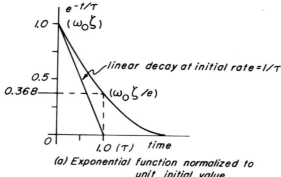

(a) Exponential function normalized to
unit initial value

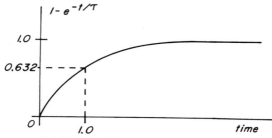

(b) Exponential rise to final value

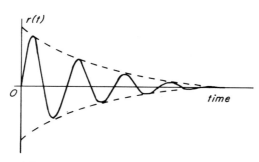

(c) Exponentially damped oscillatory response

FIG. 1.8   Responses of systems with time constants.

For electrical circuits, typical forms for τ are $\tau = RC$, $\tau = L/R$, $\tau = (2L)/R$, for R–C, R–L, and R–L–C series circuits respectively. Response functions are frequently of the form

$$r(t) = 1 - e^{-t/\tau}$$

For $t = \tau$, $f(t) = 1 - 1/e = 0.632$. The time constant for a system is

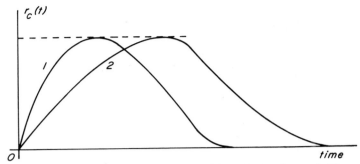

FIG. 1.9 Comparison of critically damped (1) and overdamped (2) responses.

that value of time required for response to decay to 37% of its initial value, or to rise to 63% of its final value. The critically damped response peaks and decays more rapidly than the overdamped case. In many cases, k_1 is zero and k_2 positive (Eq. 1.7). A comparison of this form of critically damped response (curve 1) and overdamped response (curve 2) is shown in Fig. 1.9.

1.5. TRANSIENT RESPONSE AND RISETIME

Experimentally, the transient response of a system can be determined by applying an input excitation V_i of the form shown in Fig. 1.10. This is called a step function input and can be produced electrically in the laboratory by a battery and switch.

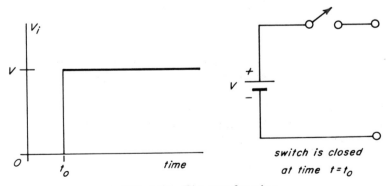

FIG. 1.10 The step function.

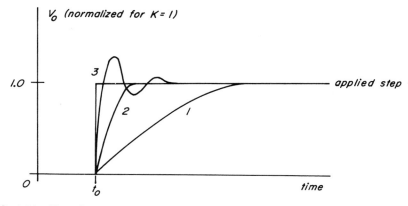

FIG. 1.11   Transient response. Typical system response to an applied step excitation. (1) Overdamped response: long response time to reach final value. (2) Critically damped response: minimum response time to final value without overshoot. (3) Underdamped response: minimum response time to final value, but oscillation ("ringing") occurs with overshooting of final value. K is a dimensionless system-gain parameter.

The output response V_0 resulting from the application of the step input is displayed on an oscilloscope or chart recorder, if V_0 is a voltage. Some typical response characteristics are shown in Fig. 1.11. Normally the response V_0 will not be identical to V_i and will take one of the forms shown in the figure.

System risetime is defined in terms of response to a step input as indicated in Fig. 1.12. There are various definitions, but usually it is taken as the time required for the output V_0 to rise from 10% of the

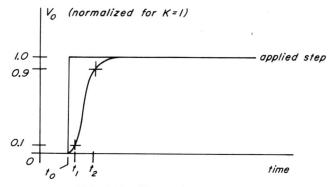

FIG. 1.12   Illustration of risetime.

final value to 90% of the final value for an applied step input V_i. With reference to Fig. 1.12,

$$\text{risetime} = t_2 - t_1$$

Referring to the transient response presented in Fig. 1.11 we see that the underdamped condition (3), yields the fastest risetime and the overdamped condition (1), the slowest risetime.

1.6.   REFERENCES

Ferris, C. D., 1962, *Linear Network Theory*, Merrill, Columbus, Ohio.
Hayt, W. H., Jr., and J. E. Kemmerly, 1971, *Engineering Circuit Analysis*, 2nd Edition, McGraw-Hill, New York.

2

Nonlinear Phenomena

2.1. INTRODUCTION

In the previous chapter, we assumed a linear system. This is only an approximation to the real physical world. Systems usually exhibit linearity only when rather small excitations are applied. Large excitations frequently produce responses that are not directly proportional to the magnitude of the excitation, resulting in nonlinear behavior.

General nonlinear-system analysis is well beyond the scope of this text because of its mathematical complexity. Several rather common nonlinear phenomena, however, can be discussed from a heuristic point of view without resorting to complex mathematical analyses. These form the subject matter of this chapter.

2.2. SIMPLE SATURATION

Figure 1.2, relating to viscous friction, illustrated a completely linear process in which the force F increases without bound as u increases. Figure 1.3 also illustrated a friction process that exhibits an initial

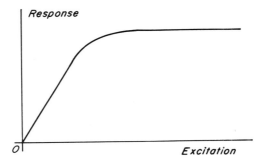

FIG. 2.1   Example of simple saturation.

linear behavior range, but then changes to a nonlinear process. A common nonlinear process is simple saturation, as depicted in Fig. 2.1. Response to the applied excitation is initially linear such that

$$r(t) = me(t)$$

where m is a constant. Over the linear range of the process

$$\frac{dr}{de} = m$$

As saturation takes effect

$$\frac{dr}{de} = f(e)$$

until hard saturation occurs when

$$\frac{dr}{de} = 0$$

Simple saturation frequently occurs in amplifiers, chart recorders, and a variety of transducers. When instrumentation systems enter saturation, calibration errors occur, components may be damaged (such as overextension of springs, mechanical linkages, and diaphragms), and performance is degraded.

2.3. HYSTERESIS

Hysteresis is normally associated with the $B–H$ properties of magnetic materials, as shown in Fig. 2.2. It also occurs in ferroelectric materials, such as the barium titanates. Hysteresis may appear in a number of

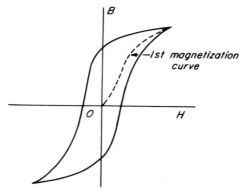

FIG. 2.2   Hysteresis in a magnetic material. B is the response magnetic field (tesla) for an applied magnetic field H (ampere-turns/meter).

components associated with an instrumentation system. The usual source is transducers fabricated from ferromagnetic or ferroelectric materials. A number of other sources, however, can be identified. Electronic trigger circuits, such as the Schmitt trigger, often possess some degree of hysteresis, as shown in Fig. 2.3. The voltage level at which the circuit turns on is different from the level at which it turns off. Thermostats and other controllers frequently exhibit hysteresis to some degree. Friction can cause hysteresis effects in springs, shafts, gears,

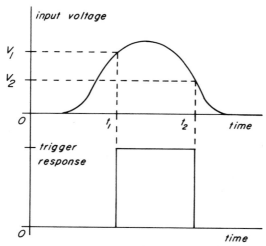

FIG. 2.3   Hysteresis in a Schmitt trigger circuit.

and mechanical linkages such that the device does not return to its original position after the excitation is removed.

This phenomenon leads to erratic system operation. It can be reduced by careful system design and maintenance (lubrication of mechanical components, etc.), but there is no convenient mathematical treatment.

2.4. DEADBAND

Deadband occurs most frequently in mechanical and electromechanical systems. It manifests as a lack of response for a change in excitation until the excitation change achieves a certain magnitude. A typical example is two shafts coupled by toothed gears. If the gears do not mesh properly (large gap between teeth), the exciting shaft can be rotated through some angle without motion of the response shaft. When the teeth finally mesh, after the initial shaft has rotated through some critical angle, then motion of the second shaft occurs. Figure 2.4 illustrates this situation.

Deadband is characterized by a delay in response and a threshold relative to applied excitation. As with hysteresis, it is not amenable to simple mathematical treatment.

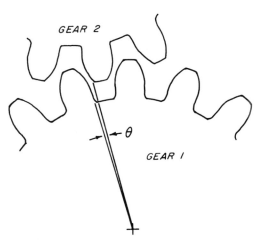

FIG. 2.4   Example of deadband between two gears. Gear 1 must move through the angle θ before it engages gear 2.

2.5. HARMONIC DISTORTION

Another form of nonlinear distortion occurs in systems that are excited by harmonic waveforms. In a linear system, the harmonic frequencies present in the response signal are the same as those in the excitation, or are reduced in number if signal filtering has taken place. In systems in which certain forms of nonlinearity exist, frequencies will be present in the response that are not present in the excitation. This is called harmonic distortion. When harmonic distortion is present, a number of undesirable system response problems can develop. Saturation leads to harmonic distortion as well as to response amplitude limiting, as discussed above.

2.6. REFERENCES

Cunningham, W. J., 1958, *Introduction to Nonlinear Analysis*, McGraw-Hill, New York.

Ku, Y. H., 1958, *Analysis and Control of Nonlinear Systems*, Ronald Press Co., New York.

Millman, J., and H. Taub, 1965, *Pulse, Digital, and Switching Waveforms*, McGraw-Hill, New York.

Section 2

TRANSDUCERS

3

Principles of
Transducer Operation

3.1. INTRODUCTION

The word transducer is derived from the Latin (*trans + ducere*, to lead across or convert). Hence, a transducer is a device used for converting input energy in one form to output energy in another form. We will be concerned with a specific class of device in which the output is an electrical quantity. The input energy may be of biological, chemical, electrical, mechanical, thermal, or optical origin.

There are three major considerations in the use of transducers. These are: 1. The transducer itself—the type of transducer that produces maximum efficiency in a particular application; 2. Detection of the electrical output from the transducer; and 3. Calibration of the electrical output in terms of the input.

The nature of the electrical output generally falls into one of the following categories: 1. A voltage (or current) produced at the output terminals of the device that is in some manner proportional to the intensity of the input disturbance; 2. A change in the electrical resistance (or impedance) of the device; and 3. A change in the electrical capacitance or inductance of the device. The method by which the output is detected is of paramount importance. In many cases, the output voltage is extremely small (on the order of microvolts or millivolts), and amplification of the signal is required before a conventional voltmeter-detector can be used. An inherent problem is electrical noise in the transducer system, so that one must design amplifiers which have low noise figures. Careful shielding of leads is essential. When the transducer output is in the form of a change in resistance, capacitance, or inductance, a bridge may be used as the detector. There are two possible modes of operation in this case. The bridge, with the transducer as one arm, may be balanced for the condition of zero input and thus rebalanced as the input develops to yield an exact value of resistance, capacitance, or inductance for the specific transducer element. On the other hand, the bridge may be balanced for the zero-input mode without any subsequent adjustment. A change in the input to the device is thus reflected as a change in the voltage across the detector arm of the bridge. The first method is useful under static conditions in which the input changes in discrete steps and sufficient time is available for balancing the bridge. The second method is to be preferred for a dynamic measurement in which the input is constantly changing and the electrical output is to be monitored by an oscillograph or strip-chart recorder.

Another technique is frequently employed to detect change in capacitance and inductance. The transducer forms part of the tuned circuit of an oscillator. A change in capacitance or inductance causes a change in the frequency of oscillation of the oscillator. The change in frequency may be detected directly by means of a frequency counter, or it may appear as a voltage that is proportional to the transducer input if an autodyne oscillator–detector is used.

In this chapter we will explore the various classes of electrically based transducers used in biomedical work. We will approach this subject based upon the mechanism by which the device works rather than the measurement application. Only fundamentals will be presented at this time; in subsequent chapters, it will then be shown how several devices may be applied to measure a particular quantity such as displacement, pressure, etc.

3.2. VARIABLE IMPEDANCE DEVICES

Operation of a number of transducers is based upon a change in electrical impedance as a function of the input excitation. In order to facilitate detection of the excitation, devices are usually designed such that variation occurs in only one of the three electrical parameters: resistance, capacitance, or inductance.

We recall that the electrical impedance of a series RLC circuit is given by

$$R + j(\omega L - 1/\omega C)$$

and that of a parallel RLC circuit by

$$\frac{R}{1 + j(\omega RC - R/\omega L)}$$

where ω is the radian frequency of the sinusoidal excitation.

3.2.1. Variable Resistance Devices

Variable resistance transducers may be divided into four general classes: 1. Those which change in resistance as a function of change in physical dimensions; 2. Those which change in resistance as a function of temperature; 3. Sliding contact devices; and 4. Semiconductor devices.

The resistance of a uniform geometry conductor is given by the expression

$$R(\text{ohms}) = \rho l / A$$

where ρ = specific resistance in ohm-meters, l = length in meters, A = cross-sectional area in square meters. If one of the physical dimensions changes, then R will change, as can be shown by finding the differential of R.

$$dR = \frac{A(\rho\, dl + l\, d\rho) - \rho l\, dA}{A^2}$$

Let us now write

$$A = k\delta^2$$

where k is a proportionality constant and δ is a sectional dimension. If A represents a square, $k = 1$; if A represents a circle, $k = \pi/4$.

Thus

$$dA = d(k\delta^2) = 2k\delta\, d\delta$$

and

$$dR = \frac{(k\delta^2)(\rho\, dl + l\, d\rho) - 2\rho l\delta\, d\delta k}{(k\delta^2)^2}$$

Now, for rectangular cross-section

$$\frac{dR}{R} = \left(\frac{A(\rho\, dl + l\, d\rho) - \rho l\, dA}{A^2}\right)\left(\frac{A}{\rho l}\right) = \frac{dl}{l} + \frac{d\rho}{\rho} - \frac{dA}{A}$$

$$= \frac{dl}{l} + \frac{d\rho}{\rho} - \frac{2\, d\delta}{\delta}$$

or

$$\frac{dR/R}{dl/l} = 1 + \frac{d\rho/\rho}{dl/l} - \frac{2\, d\delta/\delta}{dl/l}$$

or

$$\frac{dR/R}{\epsilon_a} = 1 + \frac{d\rho/\rho}{\epsilon_a} - 2\frac{\epsilon_L}{\epsilon_a} = 1 + \frac{d\rho/\rho}{dl/l} + 2\mu$$

where ϵ_a = axial strain, m/m; ϵ_L = lateral strain, m/m; μ = Poisson's ratio. The resistance strain gauge is a transducer which utilizes this principle. The factor $(dR/R/\epsilon_a)$ is known as the gauge factor for a resistance strain gauge. Strain gauges may be fabricated in a number of configurations. Some of these will be treated in the next chapter, which discusses applications of transducers.

The resistance of an electrical conductor varies with temperature according to the relation

$$R = R_0(1 + \alpha T + \beta T^2)$$

where R_0 is taken to be the resistance at 0°C, T is temperature in centigrade degrees, α is the temperature coefficient of resistivity, and α is positive for metals and negative for nonmetals. β is a second-order coefficient. Thus, a metal has a positive temperature coefficient of resistance and its resistance increases with temperature, while a nonmetal has a negative temperature coefficient and resistance decreases with temperature. This property is quite useful in temperature measuring devices.

Thermistors and resistance thermometers typify devices whose resistance changes as a function of temperature. Resistance thermometers are usually fabricated from a coil of platinum or copper wire. They are reasonably linear devices and obey rather closely the relation

$$R = R_0(1 + \alpha T)$$

over a fairly wide temperature range.

Thermistors are generally fabricated from semiconductor materials

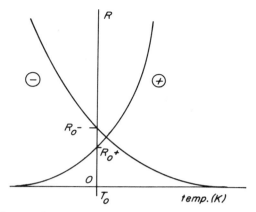

FIG. 3.1 Dependence of resistance upon temperature for thermistors (negative and positive coefficients of resistivity with temperature). Normally thermistors are operated in the negative temperature coefficient range of their characteristic curves, although barium titanate "posistors" exhibit positive temperature coefficients.

and are inherently nonlinear. They obey fairly closely linear law behavior over narrow temperature ranges, but behave exponentially over wide temperature ranges, as shown in Fig. 3.1. The exponential law behavior is not necessarily simple and is approximated by

$$R = R_0 e^{\beta(T^{-1} - T_0^{-1})}$$

in ohms at temperature T deg K, where R_0 is the resistance at some reference temperature, T_0 deg K and β is the temperature coefficient in deg K. A thermistor may be either an intrinsic or impurity semiconductor operated in a temperature range over which the conductivity changes strongly with increasing temperature.

A very simple variable resistance device is depicted in Fig. 3.2. In Fig. 3.2(a), the mechanical motion of the sliding contact changes the

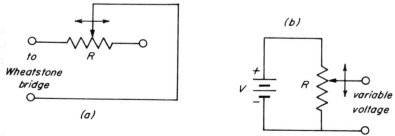

FIG. 3.2 Simple variable resistance devices: (a) unbiased; (b) biased.

resistance presented to one arm of a Wheatstone bridge. In (b), mechanical motion of the sliding contact produces a varying voltage. Such a device can be used to measure mechanical displacement and it finds application in liquid level regulators.

3.2.2.   Variable Capacitance Devices

The simplest form of capacitor consists of two parallel metallic plates of equal area separated by a dielectric material. The capacity of the capacitor so formed is a function of the geometry of the configuration and the dielectric constant of the dielectric.

$$C\text{(farads)} = \frac{\epsilon A}{l} = \frac{\epsilon_0 \epsilon_r A}{l}$$

where ϵ_0 is the permittivity of free space (8.85×10^{-12} F/m), ϵ_r is the dielectric constant (a number), A is the area of one plate (m²), l is the separation of the plates (m). Capacitive transducers generally relate a change in capacitance to a change in electrode separation l. For the coaxial geometry shown in Fig. 3.3(b)

$$C = \frac{2\pi\epsilon_0\epsilon_r l}{\ln (b/a)}$$

where a and b are the radii of the inner and outer conductors, respectively.

There are some rather serious problems regarding linearity of variable capacitance devices. If we look at a simple parallel-plate variable capacitor in which the plates are separated linearly, we can develop the following: Letting C_0 be the initial capacitance for a plate separation of l_0, then

$$C_0 = \epsilon_0\epsilon_r A/l_0$$

Letting C_1 be the final capacitance when the plates are separated to a distance l_1, in which $l_1 > l_0$ and $C_1 < C_0$, then

$$C_1 = \epsilon_0\epsilon_r A/l_1$$

$$\Delta C = C_0 - C_1 = \epsilon_0\epsilon_r A\left(\frac{1}{l_0} - \frac{1}{l_1}\right) = \epsilon_0\epsilon_r A\left(\frac{l_1 - l_0}{l_1 l_0}\right)$$

Letting $\Delta l = l_1 - l_0$, then

$$\Delta C = \frac{\epsilon_0\epsilon_r A \Delta l}{l_1 l_0} = \frac{\epsilon_0\epsilon_r A \Delta l}{l_0(l_0 + \Delta l)} = \frac{C_0 \Delta l}{l_0 + \Delta l}$$

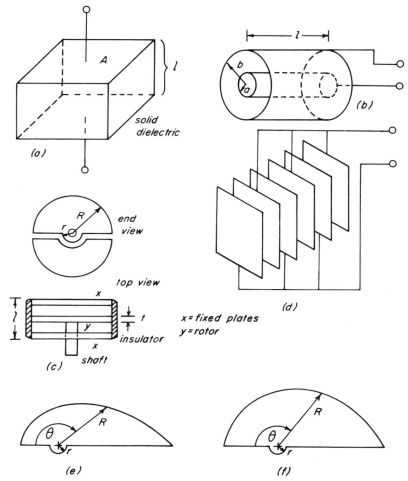

FIG. 3.3 Capacitor configurations: (a) parallel plate with solid dielectric; (b) cylindrical coaxial; (c) variable semicircular plate for linear change in C with angle θ; (d) parallel multiplate capacitor; (e) plate shape for variable capacitor which yields linear change in frequency with angle θ; (f) plate shape for variable capacitor which yields linear change in wavelength with angle θ.

For the condition of very small displacements ($\Delta l \ll l_0$), we have a linear approximation

$$\Delta C \sim C_0 \Delta l / l_0 \quad\text{and}\quad \Delta C / \Delta l \sim C_0 / l_0$$

but there is low sensitivity because of the l_0 in the denominator. In general, the capacitance of a parallel plate capacitor changes as a hyperbolic function as the plates are separated axially (simple translation). In general

$$\frac{dC}{dl} = -\frac{\epsilon_0 \epsilon_r A}{l^2}$$

To achieve a linear change in capacitance for simple translational motion, special geometries are required. Linearizing electronic circuits, as discussed below, have also been proposed.

Linear relationships can be obtained in variable capacitance devices when one electrode is rotated relative to a fixed electrode. This is the basis for the tuning capacitors used in radio and television circuits. The shape of the electrodes determines what sort of linear relation is obeyed. If semicircular electrodes are used in the configuration shown in Fig. 3.3(c), then a very nearly linear change in capacitance occurs for a linear increment in the angle of rotation, that is

$$\Delta C / \Delta \theta = \text{constant}$$

Because of the geometry, there are only 180° of effective rotation, rather than 360°. Fringing effects produce some nonlinearity at the extremum points (0–10°, 170–180°). At 0°, when the plates do not "shadow" one another, the capacitance is not zero. Greater sensitivity can be obtained by paralleling additional plates on a common rotating shaft and the fixed support.

For an array of n plates, as shown in Fig. 3.3(d), the capacitance is given by

$$C = \frac{\epsilon_0 \epsilon_r A (n - 1)}{l}$$

If a semicircular parallel-plate capacitor is constructed with one rotor plate and two fixed (stator) plates, the capacitance is

$$C = \frac{\epsilon_0 \epsilon_r (\pi/2)(R^2 - r^2)(3 - 1)}{(1/2)(l - t)}$$

$$= \frac{2\pi \epsilon_0 \epsilon_r (R^2 - r^2)}{l - t} \qquad \text{for the plates fully meshed}$$

where t is the thickness of the rotor plate, l is the separation between the two stator plates, and $(\pi/2)(R^2 - r^2)$ is the effective area of the rotor plate.

When a variable capacitance device forms part of the tuned circuit of an oscillator, as discussed in subsequent chapters, it is sometimes desirable to have a linear relation between frequency of oscillation f or wavelength λ as the angle of rotation θ changes. This can be accomplished by changing the shapes of the plates (Fig. 3.3) according to the design formulas shown below.

For a resonant circuit:

$$f = 1/2\pi\sqrt{LC} \qquad \lambda = 2\pi c\sqrt{LC}$$

where c is the velocity of light.

For linear frequency design ($\Delta f/\Delta\theta$ = constant),

$$R = \sqrt{114.6\left[\frac{2ka}{(a\theta + b)^3} + K\right]}$$

where $a = 1/180[(1/\sqrt{C_r}) - b]$, $b = C_{max}^{-1/2}$, k = (total plate area − $180K)/(C_{max} - C_r)$, $K = r^2/114.6$, C_{max} = maximum capacitance (plates fully meshed), C_r = residual capacitance (plates fully unmeshed), r = radius of shaft-rotor plate joint, R = plate radius, and $C = (a\theta + b)^{-2}$.

For linear wavelength design ($\Delta\lambda/\Delta\theta$ = constant)

$$R = \sqrt{114.6[2kx(x\theta + y) + K]} \qquad C = (x\theta + y)^2$$

where $x = (\sqrt{C_{max}} - \sqrt{C_r})/180$, $y = \sqrt{C_r}$, and the other constants are as above.

If fixed capacitors are connected either in series or in parallel with these variable capacitors, the linearity relations are modified, since both C_r and C_{max} are changed. For parallel connection, C_r and C_{max} are increased by the same fixed amount (the value of the fixed parallel capacitor). For series connection of a fixed capacitor C_s, the equivalent values C'_{max} and C'_r are

$$C'_{max} = C_s C_{max}/(C_s + C_{max}) \qquad \text{and} \qquad C'_r = C_s C_r/(C_s + C_r)$$

The Wayne Kerr Corporation (Philadelphia, Pa.) has developed a feedback system to linearize the nonlinear capacitance vs. displacement characteristic for translational capacitive transducers. An operational amplifier (Chapter 7) is used in a feedback mode, as shown in Fig. 3.4. Two conditions need to be satisfied: 1. The amplifier input impedance must be very large so that input signal current can be neglected. 2. The amplifier gain must be very high such that when its output is not saturated, the input voltage v_i is essentially zero. These conditions can be

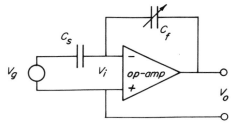

FIG. 3.4   Circuit for linearizing linear displacement variable capacitance transducer. Op-amp bias connections are not shown.

satisfied with most modern integrated circuit operational amplifiers. In a practical system, the external generator voltage v_g is a sine wave of constant amplitude ($V_g \sin \omega_0 t$).

When the operating conditions are satisfied, the output voltage v_0 is (Doebelin, 1975)

$$v_0 = -\frac{C_s l v_g}{\epsilon_0 \epsilon_r A}$$

where C_s is the fixed series capacitance, l is the plate separation of the transducer (C_f), and A is the plate area. Thus the output voltage v_0 is a sine wave, amplitude modulated by the mechanical motion of the transducer. For rest conditions when $\Delta l = 0$ and $l = l_0$,

$$v_0 = -\frac{C_s l_0 v_g}{\epsilon_0 \epsilon_r A}$$

This is an unmodulated "carrier" wave. If the plate separation is varied sinusoidally as in a vibration sensor, then

$$l = l_0 + \Delta l \sin \omega t$$

and

$$v_0 = -\frac{C_s(l_0 + \Delta l \sin \omega t)v_g}{\epsilon_0 \epsilon_r A} = -\frac{C_s(l_0 + \Delta l \sin \omega t)V_g \sin \omega_0 t}{\epsilon_0 \epsilon_r A}$$

To achieve proper carrier modulation, ω and ω_0 must be widely separated in frequency ($\omega_0 \gg \omega$). The output voltage is then processed in an envelope detector (Chapter 8). The signal may then be displayed on an oscilloscope or chart recorder as a dynamic record, or it may be rectified and filtered (Chapter 8) to obtain a DC voltage that is proportional to maximum displacement.

Another linear displacement scheme involves moving a dielectric between two fixed capacitor plates. Recall that for a parallel plate capacitor

$$C = \frac{\epsilon_0 \epsilon_r A}{l}$$

where $\epsilon_r = 1$ for air. If ϵ_r varies, then

$$\frac{dC}{d\epsilon_r} = \frac{\epsilon_0 A}{l}$$

Thus we can build a linear displacement transducer by moving a solid dielectric slab or a liquid between two fixed metal plates. The slab is connected via a mechanical linkage to the moving object. The major drawback of this scheme is that practical dielectrics have an $\epsilon_r \sim 2.0$–2.5. For the displacements normally encountered in biomedical work, ΔC would be quite small. It is not practical to attempt to vary plate area, which would also give a linear relation. A possible technique, however, is to move one plate relative to a fixed plate by sliding the movable electrode past the fixed one. The effective area then changes linearly with the amount that the sliding plate is translated. Imagine that the moving plate shadows the fixed element. Assuming parallel rectangular plates and neglecting field fringing, let w be the width of the plates and x the amount that the moving plate moves. Then the effective area is $A = wx$, and

$$C = \epsilon_0 \epsilon_r wx/l \qquad \frac{dC}{dx} = \epsilon_0 \epsilon_r w/l = \text{constant}$$

This is a linear relationship for translational motion. If a multiplate configuration is used as for the rotating capacitor (Fig. 3.3), increased sensitivity can be obtained. In practice, there will be some electric field fringing so that nonlinearities will occur, as observed in rotating capacitors. These will occur at the extremum points when the plates are fully meshed and fully unmeshed. Some of the fringing may be reduced by making the fixed electrode much larger in surface area than the movable electrode.

The coaxial geometry shown in Fig. 3.3 may also be used to obtain a linear change in capacitance with displacement since

$$\frac{dC}{dl} = \frac{2\pi\epsilon_0\epsilon_r}{\ln(b/a)} = \text{constant}$$

In a practical device the inner conductor would be moved relative to the fixed outer sleeve. As with parallel plate capacitors, electric field fringing is also a problem in this geometry. As long as the major portion of the inner conductor remains surrounded by the outer sleeve, reasonably linear operation can be expected. If the inner conductor is completely extracted from the sleeve, serious nonlinear problems will develop.

3.2.3. Variable Inductance, Magnetic, and Faraday-Law Devices

Many of the devices which fall into this category are based upon Faraday's Law of Induction, which states that if relative motion occurs between an electrical conductor and a magnetic field, an electromotive force is produced in the conductor. This is the principle of operation of most electric generators. If one moves an electrical conductor such as a metallic rod so that lines of magnetic flux are cut, a voltage arises that may be measured across the two ends of the rod. It is assumed that the magnetic field is static. On the other hand, if the rod is fixed in space in the region of a time-varying magnetic field and oriented perpendicular to the field, a voltage will again be produced. A more complicated situation involves both motion of the conductor and time variation of the magnetic field. Mathematically, Faraday's Law may be stated in the following manner for a closed circuit:

$$\text{emf} = -\frac{d\phi}{dt} \tag{1}$$

where emf is in volts, ϕ is the total flux in webers, and t is the time in seconds (mks units). If there is more than one turn of wire in the circuit, then we may write

$$\text{emf} = -N\frac{d\phi}{dt} = -\frac{d\lambda}{dt}$$

where N is the number of turns linking the flux, and λ is the total flux linkage in weber-turns. The emf is found by performing a line integration of the electric field $\mathbf{E}$ around the closed path of the circuit shown in Fig. 3.5.

$$\text{emf} = \oint \mathbf{E} \cdot \mathbf{dl} = -\frac{d\phi}{dt} \qquad \text{per loop} \tag{2}$$

Now

$$\phi = \iint_s \mathbf{B} \cdot \mathbf{n}\, ds$$

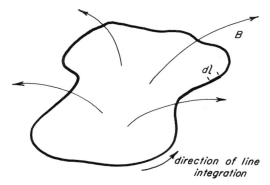

FIG. 3.5   Emf determined by line integration along circuit conductor.

and

$$-\frac{d\phi}{dt} = -\frac{d}{dt}\iint_s \mathbf{B}\cdot\mathbf{n}\,ds \qquad (3)$$

where s is the surface enclosed by the conducting loop, and $\mathbf{n}$ is the unit normal vector to the surface s. Combining Eqs. (2) and (3), we write

$$\oint \mathbf{E}\cdot\mathbf{dl} = -\frac{d}{dt}\iint_s \mathbf{B}\cdot\mathbf{n}\,ds$$

Faraday's law may now be expressed as follows:

$$\text{emf (volts/turn)} = -\frac{d}{dt}\iint_s \mathbf{B}\cdot\mathbf{n}\,ds \qquad \text{general} \qquad (4)$$

$$\text{emf (volts/turn)} = -\iint \left(\frac{\partial\mathbf{B}}{\partial t}\right)\cdot\mathbf{n}\,ds$$

$$\text{conductor fixed with time variation of field} \qquad (5)$$

The case of motion of a conductor relative to a fixed magnetic field may be handled more explicitly. Experimental observations have shown that if an electric charge q moves with a velocity $\mathbf{u}$ in a magnetic field $\mathbf{B}$, a force $\mathbf{F}$ is exerted upon the moving charge. (This is the principle used in magnetic deflection television tubes.)

$$\mathbf{F} = q(\mathbf{u} \times \mathbf{B}) \qquad (6)$$

where q is the electric charge in coulombs, $\mathbf{u}$ is the velocity in m/s, and $\mathbf{F}$ is the force in newtons. Another experimental law (from Coulomb) indicates that force, charge, and electric field intensity are related as

$$E = \frac{F}{q} \tag{7}$$

If we now combine Eqs. (6) and (7), we may write

$$E = u \times B$$

and integrate in the manner used to develop Eq. (2).

$$\text{emf (volts/turn)} = \oint E \cdot dl = \oint (u \times B) \cdot dl$$

$$\text{conductor moves relative to fixed field} \tag{8}$$

Combining Eqs. (5) and (8) we may write

$$\text{emf (volts/turn)} = \oint (u \times B) \cdot dl - \iint \left(\frac{\partial B}{\partial t}\right) \cdot n \, ds \quad \text{general case}$$

$$\tag{9}$$

If the conductor always moves in a direction perpendicular to the field lines, then we may simplify Eq. (8).

Now $|(u \times B)| = uB \sin \underline{/u, B} = uB \sin \theta$, where $\theta = 90°$. Hence $|(u \times B)| = uB$, and $(u \times B) \cdot dl = |u \times B| \, |dl| \cos \theta$, where θ, the angle between the resultant vector ($u \times B$ and dl) is $0°$ in this case. Hence $(u \times B) \cdot dl = uB \, dl$ and

$$\text{emf (volts/turn)} = \oint (u \times B) \cdot dl = \oint uB \, dl = uBl \tag{10}$$

where l is the length of the conductor which cuts the flux lines.

Faraday's Law is the basis for many devices, among which are transformers (fixed conductor with time-varying field), differential transformers (moving conductors and time-varying field), generators, tachometers, and special devices, some of which will be discussed later.

The lossless transformer may be characterized by the equations shown below.

$$\frac{V_2}{V_1} = \frac{N_2}{N_1} \qquad \frac{I_2}{I_1} = \frac{N_1}{N_2} \qquad \frac{V_1}{V_2} = \frac{I_2}{I_1} \tag{11}$$

where V and I are assumed phasor quantities, and V_1 = primary volts, V_2 = secondary volts, I_1 = primary amperes, I_2 = secondary amperes, N_1 is the number of primary turns, and N_2 is the number of secondary turns. We note that $Z_1 = V_1/I_1$; $Z_2 = V_2/I_2$, hence

$$\left(\frac{V_2}{V_1}\right)^2 = \left(\frac{N_2}{N_1}\right)^2 = \frac{V_2(Z_2 I_2)}{V_1(Z_1 I_1)} = \frac{N_2}{N_1}\frac{Z_2}{Z_1}\frac{N_1}{N_2} = \frac{Z_2}{Z_1}$$

$$\sqrt{\frac{Z_2}{Z_1}} = \frac{N_2}{N_1} \tag{12}$$

Many devices utilize a change in a magnetic circuit for their operation. Of these, many in common use depend upon a change in inductance. The voltage–current relations for an inductor are

$$v(t) = L\frac{di(t)}{dt}$$

$$i(t) = \frac{1}{L}\int v(t)\,dt \qquad \text{for constant } L$$

$$v(t) = \frac{d}{dt}(Li) = \frac{d\lambda}{dt} \qquad \text{general, from Faraday's Law}$$

where L is the inductance in henries. Expanding the last relation, we find

$$v(t) = L\frac{di}{dt} + i\frac{dL}{dt} \tag{13}$$

In the first term, inductance is fixed and current varies, while in the second term, current is fixed and inductance varies. We shall now examine the relation

$$v(t) = i\frac{dL}{dt} \tag{14}$$

Three geometries commonly found in transducer assemblies, as shown in Fig. 3.6, are the long solenoid, the toroid, and the coaxial cylinder. Inductance is defined as the rate of change of flux linkages with current.

$$L = \frac{d\lambda}{di}$$

The magnetic field B produced by a current I in a long solenoid is

$$B = \frac{\mu NI}{\sqrt{4R^2 + l^2}} \qquad \text{on coil axis at the coil midpoint}$$

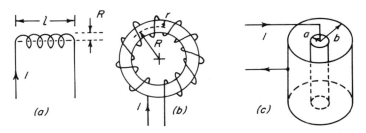

FIG. 3.6   Inductor geometries: (a) solenoid; (b) toroid; (c) coaxial cylinder.

where μ is the permeability of the core of the solenoid (henries per meter), N is the number of turns, I is the current in the solenoid in amperes, R is the radius of the solenoid in meters, and l is the length of the solenoid in meters. If $l \gg R$, then $B \approx \mu NI/l$.

For simple geometries, $\lambda = N\phi$ and $\phi = BA$, where ϕ = total flux produced by one turn; A is the area of turns (cross-section), and $L = \lambda/I$. Hence, for the long solenoid

$$L = \frac{NBA}{I} = \frac{\mu N^2 A}{l} = \frac{\mu \pi (NR)^2}{l} \text{ henries (H)}$$

for the toroid

$$L = \frac{\mu (Nr)^2}{2R} \text{ henries (H)}$$

for the coaxial cylinder

$$L = \frac{\mu l}{2\pi} \ln(b/a) \text{ henries (H)}$$

Transducers may be developed from these geometries most simply by changing the magnetic permeability μ of the core. The magnetic permeability of the core is a specific property of the material used. The permeability of air is

$$\mu_0 = 4\pi \times 10^{-7} \text{ henries per meter (H m}^{-1})$$

and the relative permeability of other materials ranges from approximately unity for nonferromagnetic materials to over one million for ferromagnetic alloys.

$$\text{relative permeability} = \mu_r = \mu/\mu_0$$

Thus, if we design a solenoid with a movable core and maintain a constant current in the coil, we can change the voltage across the coil by motion of the core, provided that there is an effective change in μ. If the core is a permanent magnet, we can produce a voltage across the coil without the requirement of a current in the coil simply by moving the magnetized core and invoking Faraday's Law.

Some transducers are based upon the principle of variable reluctance. In magnetic circuits there is a relation which corresponds to the familiar Ohm's Law of electric circuits. This may be stated (analogous to $V = RI$) as

$$\mathscr{F} = \mathscr{R}\phi$$

where $\mathscr{F}$ is the total magnetomotive force (mmf) through the magnetic circuit in amperes, ϕ is the total flux through the magnetic circuit in webers, and $\mathscr{R}$ is the total reluctance of circuit in henries^{-1}. For a simple geometry with uniform cross-section

$$\mathscr{R} = \frac{l}{\mu A}$$

Reluctance may be varied by changing the length l, area A, or permeability $\mu = \mu_r \mu_0$. The change in $\mathscr{R}$ is usually detected as a change in inductance as described above, or as a varying voltage.

The common magnetic transducer configurations are shown in Fig. 3.7. The configurations shown are the bases for many phonograph cartridges.

In closing, it should be noted that ferromagnetic materials are nonisotropic. The magnetic permeability is determined by the ratio of the **B**-field, measured in teslas, and the **H**-field, measured in amperes per meter, where $\mathbf{B} = \mu \mathbf{H}$. The permeability varies depending upon **B**

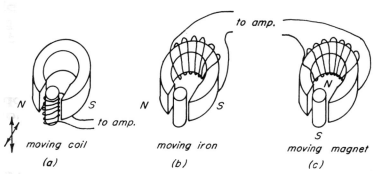

FIG. 3.7   Magnetic transducers: (a) moving coil (Faraday-Law device); (b) moving iron (variable reluctance); (c) moving magnet (Faraday-Law device).

and **H**, as shown by the hysteresis loop of Fig. 2.2. Hysteresis phenomena account for many of the nonlinear properties of magnetic transducers.

3.3. PIEZOELECTRIC AND FERROELECTRIC DEVICES

A piezoelectric material exhibits several interesting properties. If a crystal of the material is cut in the form of a parallel plate and the opposing parallel sides are silver plated with electrodes attached, and if the crystal is then subjected to a mechanical stress, a voltage can be measured across the electrodes. When a voltage is applied to the electrodes, a mechanical strain results in the crystal. In some crystals, over given temperature ranges, a permanent electric polarization exists analogous to the permanent magnetism in a ferromagnetic material. Common piezoelectric crystals are quartz, barium titanate ($BaTiO_3$), and Rochelle salt (sodium potassium tartrate, $KNaC_4H_4O_6 \cdot 6H_2O$). Barium titanate and Rochelle salt also exhibit ferroelectric properties. Some transducers use zirconates rather than titanates.

The cause of piezo- or ferroelectricity is complex and one must examine the crystal lattice structure of the material to understand completely the mechanisms involved. On the macroscopic level, a network representation for a piezoelectric crystal may be developed as illustrated in Fig. 3.8, in which Z_m is a mechanical impedance, Z_a an acoustical impedance, and K is an electromechanical coupling coefficient. For a vibrating piezoelectric crystal (steady-state AC case), these quantities may be expressed as follows:

$$Z_a = \tfrac{1}{2}(AdC_m)$$

$$Z_m = \frac{1}{2}\left(\frac{MA}{C}\right) \coth \left(\frac{j\pi h}{\lambda} + \frac{\alpha h}{2}\right)$$

$$K = \frac{AMq}{h}$$

$$C = (\epsilon_r\epsilon_0 - q^2 M)(A/h)$$

where A is the area of the plate (electrode area), M is Young's modulus, q is the piezoelectric constant (2.12×10^{-12} coulomb/newton for quartz, CN^{-1}), d is the density of the medium outside the crystal, C_m is the velocity of sound in the external medium, C is the velocity of sound in the crystal, λ is the wavelength of sound in crystal, α is the mechanical loss factor, h is the thickness of the crystal, ϵ_0 is the per-

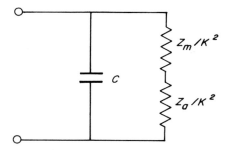

FIG. 3.8   Electric network representation for a piezoelectric crystal.

mittivity of free space ($8.85 \times 10^{-12}\ \mathrm{F\ m^{-1}}$), and ϵ_r is the relative permittivity (dielectric constant). The vibrating piezoelectric crystal transforms (transduces) the mechanical impedance $(Z_a + Z_m)$ into an electrical impedance $(Z_a + Z_m)/K$.

For the present study we are only interested in the property of the crystal by which mechanical stress is transformed into a voltage and vice versa.

The ferroelectric material derives its name from the fact that it exhibits a hysteresis loop for **D** and **E** analogous to the **B**–**H** hysteresis loop for ferromagnetic materials (see Fig. 2.2).

Depending upon geometry and initial polarization, as shown in Fig. 3.9, piezoelectric crystals can be made to vibrate in linear longitudinal, radial, or shear modes. Before they are used for the first time, the crystals must be polarized using a high DC voltage.

3.4. THERMOELECTRIC DEVICES

There are three thermoelectric phenomena which are of interest. These are the Peltier, Thomson, and Seebeck effects. If two dissimilar metallic electrical conductors are joined together at a point, and the junction is held at constant temperature while a current is passed through the circuit, heat is absorbed or generated at the junction in a quantity additional to the i^2R Joule heat generated. This is the Peltier effect. If a current exists in a conductor along which there is a thermal gradient, then heat is generated or absorbed in addition to the i^2R Joule heat produced, a property known as the Thomson effect. The Thomson effect is much smaller than the Peltier effect. Peltier-effect devices are frequently used to cool samples in osmometers designed to measure total electrolyte base.

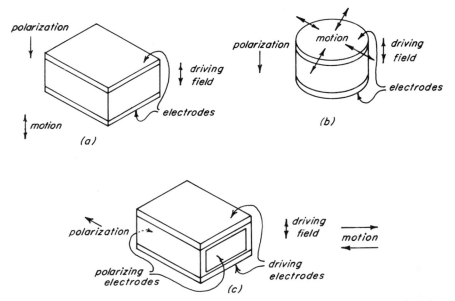

FIG. 3.9 Vibratory modes of ceramic transducers: (a) thickness mode of a ceramic plate; (b) radial mode of a ceramic disc; (c) thickness shear mode of a ceramic plate.

If two dissimilar metallic conductors are joined at a point and the junction heated, then an emf is produced in the circuit, a phenomenon known as the Seebeck effect. The Peltier and Thomson effects have applications in various thermoelectric heating and cooling devices, while the Seebeck effect is the basis for thermocouple action.

Thermocouples formerly found extensive use in biomedical work where accurate temperature measurements were required. In many applications, they have now been replaced by thermistors. They are still useful for high temperature measurements where thermistors would be destroyed by the heat involved.

3.5. PHOTOELECTRIC DEVICES

Many transducers depend upon photoelectric phenomena. Some convert light energy into electrical energy directly without requiring applied energy from other sources. The cadmium sulfide solar battery is an example of such a device. Others, such as the conventional photo-electric cell, require an electrical bias in order that an efficient conversion

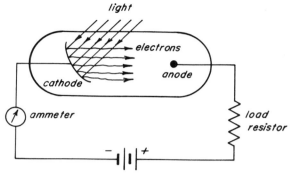

FIG. 3.10   Biased photoelectric device.

from light to electrical energy take place. Although there are numerous photo effects which occur in nature, one may explain the photoelectric effect as follows: Let us suppose that we have a metal or metal oxide surface. Light impinges upon the surface and electrons are produced which are detected as a current (see Fig. 3.10). The maximum energy of the electrons produced is found to be proportional to the intensity of the incident light as expressed by the Einstein–Planck relation

$$E_{max} = h\nu - E_0$$

where E_{max} is the maximum energy of an emitted electron, E_0 is an energy associated with the emitting surface, ν is the frequency of incident light, and h is Planck's constant, a universal constant. It can be derived from the Einstein–Planck relation stated above that electrons are emitted in discrete energy steps or quanta. There are two such distinct photo-emission effects in a metal. One is a surface effect, whereas the other is a volume effect. In the surface effect, an impinging light photon re-leases an electron at the metal surface. In the volume effect, the photon is absorbed in the interior of the metal, in which its energy is imparted to a lattice electron that eventually makes its way to the surface and escapes.

Two effects exist in semiconductor junction diodes. The photo-electric effect occurs when a hole-electron pair is produced by an impinging photon. The diode requires bias for this mode of operation. If the diode is unbiased and the junction is left open-circuited, a voltage that is proportional to the incident photon energy is produced across the junction; this is known as the photovoltaic effect.

The effects described above may be utilized in various transducers to produce light detectors, both general and frequency sensitive, as well as energy conversion devices to produce an electrical output for an applied input.

There are numerous photoelectric devices. The more commonly used light sensors are photodiodes, phototransistors, photomultiplier tubes, and photoresistors.

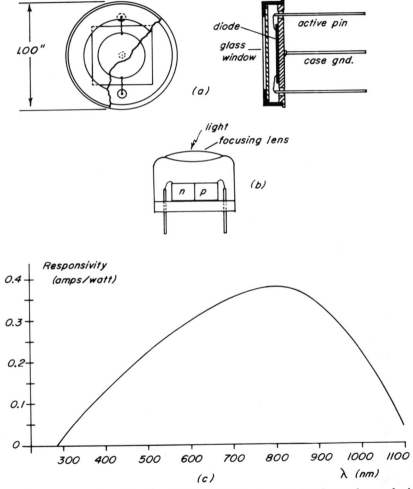

FIG. 3.11 Silicon photodiodes: (a) United Detector Technology planar device (PIN-10DB/541); (b) mechanical design of typical general purpose silicon photodiode; (c) typical spectral response of silicon photodiode (United Detector Technology device). Figures (a) and (c) redrawn, with permission, from United Detector Technology, Inc. Specification Sheet D-014-0267.

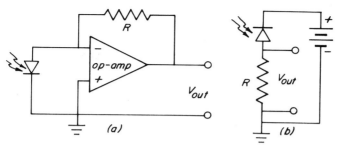

FIG. 3.12   Biasing arrangements for silicon photodiodes: (a) United Detector Technology devices; (b) typical general purpose silicon devices. Op-amp bias connections are not shown.

Normally, semiconductor photodiodes are doped silicon *pn* junctions operated in the reverse bias mode. Figures 3.11(a) and (b) illustrate the basic construction of these devices, while Fig. 3.12 describes two biasing circuits. Spectral sensitivity, as shown in Fig. 3.11(c), is a function of the silicon material, but is basically the same for all silicon photodiodes. The maximum sensitivity is in the near IR at approximately 800 nm. A more flat response can be obtained by using an appropriate optical filter in front of the diode, at the expense of overall sensitivity. Narrow bandwidth filter lenses (red, green, etc.) can be used to obtain peak sensitivity for particular wavelength bands. It must be recognized, however, that this will not increase sensitivity. For example, if a filter lens that has maximum transmission at 500 nm is used, then from Fig. 3.11(c) we can see that the *maximum* responsivity is 0.22 A/W. In silicon photodiodes, photocurrent is linearly proportional to impinging light intensity.

Because of the spectral sensitivity limitations of silicon photodiodes, phototubes are still used in many applications. Several designs are illustrated in Fig. 3.13(a). Volt–ampere characteristics and spectral responses for several photocathodes are shown in Figs. 3.13(b) and (c). Biasing is shown in Fig. 3.14. Phototubes consist of a curved light-sensitive cathode and a narrow cylindrical anode. The glass bulb may contain a vacuum or be filled with an inert gas, such as argon, under very low pressure. Light (photons) striking the cathode produces electrons which are collected at the anode. This generates the plate current shown in Fig. 3.13(b). The S-1 surface [Fig. 3.13(c)], which responds over the visible light range, is a composite of silver, cesium, and cesium oxide.

There are several phototransistor configurations. One, shown in

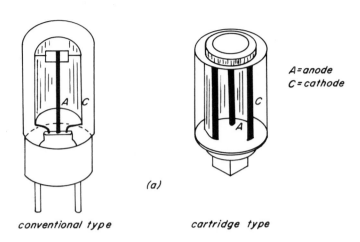

A = anode
C = cathode

conventional type   cartridge type

(a)

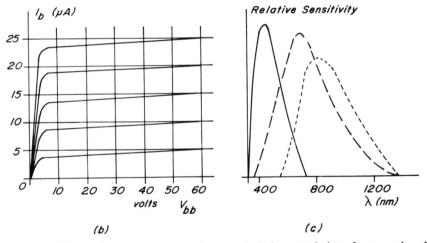

(b)   (c)

FIG. 3.13 Phototube configurations: (a) physical characteristics of conventional and cartridge type phototubes; (b) typical volt–ampere characteristics for anode current (I_b) as a function of anode supply voltage (V_{bb}); (c) relative spectral sensitivities for photosensitive cathode surfaces; reading from left to right, the curves approximate standard surfaces S-4, S-3, S-1.

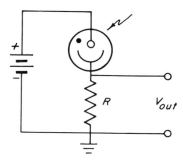

FIG. 3.14   Biasing arrangement for phototube. The black dot denotes a gas-filled device (frequently argon under low pressure).

Fig. 3.15, is basically a resistance-type cell with a photoemitter. The active leads are the base and collector. Typical characteristcs are:

V_{CB}	25–40 V DC
I_C	1–3 mA DC
$P_{diss.}$	200 mW max.
Internal resistance	20 kΩ
Power sensitivity	0.5 mW/mlumen

Other devices have the normal three transistor connections, but in many applications they are operated in the reverse bias mode with the base unconnected, as shown in Fig. 3.16. Typical maximum operating characteristics for these devices (Monsanto MT1 and MT2, for example) are:

BV_{CEO}	30 V DC
BV_{ECO}	7 V DC
BV_{CBO}	80 V DC
I_C	40 mA
$P_{diss.}$	200 mW
Sensitivity, S_{CEO}	200–1400 μA/mW/cm^2 (depending upon device) at $\lambda = 900$ nm

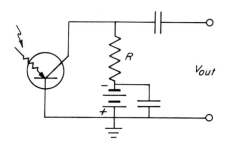

FIG. 3.15   Biasing arrangement for resistive-type phototransistor.

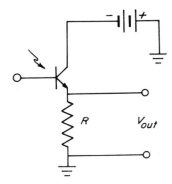

FIG. 3.16   Biasing arrangement for silicon phototransistor.

Although photodiodes and phototransistors are rather sensitive to light, they are not sufficiently sensitive for some applications, such as those involving scintillation counters and many spectrophotometers. For these applications, photomultiplier (PM) tubes, described in Fig. 3.17, are used. The device is a photodiode in conjunction with a series of electron multiplication electrodes (dynodes). In theory, the electrons emitted by the photocathode are accelerated through a potential difference, thereby gaining momentum (kinetic energy). They collide inelastically with the low-work-function first dynode, thus producing secondary electrons in a quantity greater than those initially impacting this dynode. This process is repeated between dynodes so that effective photocurrent multiplication occurs. Typical operating characteristics are shown below. These devices are available in a number of spectral characteristic types. Current amplification factors of 10^6 are typical.

Spectral response	S-11
Wavelength of maximum response	4400 nm
Cathode, semitransparent	
shape	circular
Window	
Area	14.2 cm²
Minimum diameter	4.3 cm
Index of refraction	1.51
Direct interelectrode capacitances (approx.)	
Anode to dynode No. 10	6.0 pF
Anode to all other electrodes	2.4 pF
Overall length	17.25 cm
Seated length	14.98 ± 0.33 max. cm
Maximum diameter	5.72 cm
Bulb	T-16
Base	small-shell diheptal

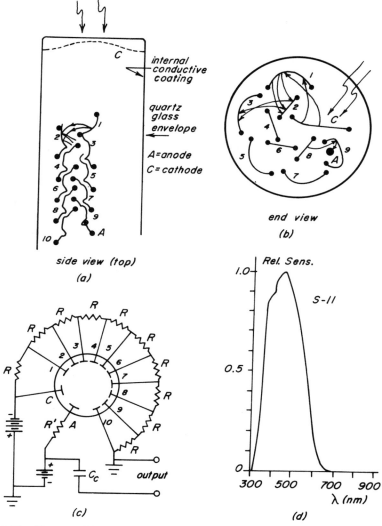

FIG. 3.17 Photomultiplier tube characteristics: (a) mechanical design of a Western Electric Code BC 10-stage photomultiplier; (b) mechanical design of a typical photomultiplier tube; (c) biasing circuit for a 10-dynode photomultiplier tube; (d) S-11 spectral characteristic of the Western Electric device. Note that the sensitive surface is the semitransparent photocathode indicated as "*C*" in (a).

Maximum ratings, absolute values
    Anode supply voltage   2000 max. V DC
    Supply voltage between dynode No. 10 and anode   400 max. V DC
    Supply voltage between cathode and dynode No. 1   400 max. V DC
    Supply voltage between cathode and dynode No. 3   800 max. V DC
    Focusing-electrode voltage   400 max. V DC
    Average anode current   2 max. mA
    Ambient temperature   75 max. °C

Photoresistors (photoconductors), frequently called photocells (this name has also been applied to phototubes) are generally amorphous deposits of materials such as Cds, CdSe, or CdTe. Basically these devices are resistors whose resistance changes as a function of applied light. Figure 3.18 illustrates general schematic and physical characteristics for these devices. CdS cells find application in camera light meters and light-operated relays. Generally speaking, these devices are less sensitive than the other devices discussed, although some are highly sensitive. Their main drawbacks are slow response time and high noise level. All photoelectric devices produce some electrical noise, but this is especially high in photoresistors. They are also more sensitive to temperature than other devices. There is a tradeoff between sensitivity and response time: the faster the response, the lower the sensitivity to applied light.

Another consideration in selecting photodetectors is dark current. This is the residual current in the device (analogous to I_{co} of a transistor) when the device is totally shielded from light. Compared to other devices, dark current is high in phototubes, especially those which are

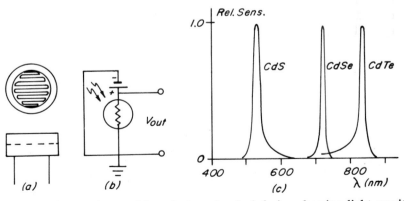

FIG. 3.18 Photoresistors: (a) typical mechanical design showing light sensitive grid; (b) biasing circuit; (c) spectral characteristics for three different chemical compositions.

gas filled. In phototubes and PMs, it increases with the age of the device until it reaches unacceptable levels and the device must be replaced.

3.5.1. Light Sources

Many light sources are used in instrumentation systems. The simplest is the tungsten incandescent filament. This is simply an ohmic resistor which emits light (IR-visible) as a function of applied current. When particular spectral characteristics are required, gas discharge devices are used. These contain two separated electrodes in a gas-filled envelope. The gas is ionized by a voltage applied across these electrodes. Basic operation is governed by Paschen's Law, which states:

$$V_s = \text{constant}(pd)$$

where V_s is the minimum voltage required to produce discharge, p is the gas pressure, and d is the electrode separation. The value of the constant depends upon the gas or gas mixture used. To some extent, V_s is also a function of geometry and electrode material. Typical gases used are argon, deuterium, hydrogen, mercury vapor, neon, sodium vapor, and xenon. Normally, operation is in the arc-discharge region rather than the glow-discharge range.

For narrow-wavelength band operation, light emitting diodes (LEDs) find many applications. In operation, they are biased in the forward direction using a series current-limiting resistor for protection, as shown in Fig. 3.19. They may be operated from DC, AC, or pulsed sources. As current passes through the diode, hole-electron pairs are produced. When recombination occurs, the lost energy is radiated as light. Typical materials and the wavelengths of the light emitted are:

GaP (Zn, O-doped, red)	698 nm
GaP (Zn diffused, green)	553 nm
$Ga_{1-x}Al_xAs$ (red)	688 nm

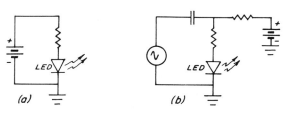

FIG. 3.19   LED biasing circuits.

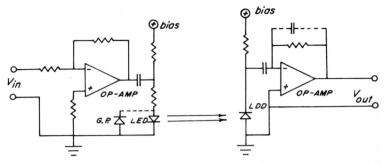

FIG. 3.20   Possible realization for electro-optical isolator. Op-amp bias connections are not shown.

Typical LED characteristics are, at 25°C:

Forward voltage	2.6 V
Luminous intensity	4.5 mcd at 40 mA
Reverse current	0.001 μA at 4 V
Reverse voltage (max.)	4 V
Forward DC current (max.)	40 mA
Power dissipation (max.)	225 mW

Lasers are also used in many applications as light sources. Their theory of operation is beyond the scope of this text. The interested reader is referred to the volume by Siegman, 1971.

Optical isolators are used in a number of instrumentation systems when it is necessary to isolate completely two segments of a circuit or system. Figure 3.20 illustrates one method for implementing this device. Normally, the LED and LDD would be placed at opposite ends of a light-tight tube. Optical isolators are also available in commerical packages. In operation, a signal (current) drives the LED to produce a light signal directly proportional to the electrical drive. Coupling is through the light tube to an LDD. The output of the LDD is then amplified. One application of this technique is discussed in section 12.2. Matched linear LEDs and LDDs must be used. Operation is normally in the red wavelength range. The LED must be statically forward-biased, as shown in the figures, so that the signal does not drive it into the reverse-bias range with resultant clipping and possible damage to the LED. A protective general purpose diode may be used in addition as indicated by the dotted connection in the figure.

3.6.   FIBER OPTICS

Fiber optic techniques are relatively new and they are currently finding numerous applications in medical equipment, as well as in communications systems and industrial instruments. A primary medical application is endoscopy, that is, the illumination and optical probes inserted into the natural body openings for various types of clinical evaluations.

A fiber optic probe is essentially a light tube or optical waveguide analagous to the waveguides used in microwave systems. They are sometimes called dielectric waveguides. Two basic designs exist. One employs a light pipe or light tube to conduct light from one location to another. In medical applications, it is used to bring light from an external light source to an internal portion of the body, such as the oral cavity, colon, etc. It is solely an illumination transducer. The other design, which consists of an array or bundle of light tubes, is used for imaging, that is transmission of an optical image from one point to another, as discussed subsequently.

A single optical fiber generally consists of a fine glass rod of high refractive index with a surrounding sheath of low refractive index glass, hence a coaxial glass tube. For some purposes, the fiber is coated with a thin layer of metal, such as aluminum. Single fibers frequently have diameters on the order of 0.003 in. or 50–60 microns. A typical configuration consists of a central core of flint glass with a refractive index of 1.62 sheathed with a glass of refractive index equal to 1.52. Because of the difference in refractive indices, total reflection occurs at the interface between the two glasses. Thus any light introduced into the center core remains in the core; none escapes through the sheath. This maintains light intensity and prevents "cross-talk" when fibers are incorporated into a bundle.

These single fibers may be gathered into cylindrical bundles of as many as 600 individual fibers. The bundle ends are treated in various manners depending upon subsequent application. For imaging, they are frequently machined to have planar surfaces perpendicular to the axis of the fibers. Typical bundle diameters are 0.03–0.25 in. Bundles are often enclosed by a PVC sheath which increases overall diameter to 0.187–0.32 in.

An endoscopic probe may consist of two bundles, one for illumination and one for imaging. Light is brought to an internal body site by the illumination portion of the probe. The imaging portion of the probe transmits an image of the illuminated area to an external viewing mechanism or eyepiece. Probes have also been developed for transmission of high intensity laser energy for certain surgical applications.

Aside from the imaging aspect, fiber optic bundles permit high intensity illumination without concomitant heating, because the light source is remote from the illuminated site. They are flexible and small in diameter. More information on general applications and fabrication is to be found in Lisitsa et al., 1972, and Holliday and Stow, 1968.

3.7.   SUMMARY

The variety of transducers is virtually limitless. The phenomena described in this section are those which are most frequently used in measurement applications in both physical and biological systems. Many effects have not been mentioned, such as change in the electrical resistance of a material under an applied magnetic field (Hall effect), and change in a material's optical properties as a function of an applied electric field (Kerr effect) or an applied magnetic field (Faraday effect). Some of these will be discussed in subsequent chapters.

3.8.   REFERENCES

Beckwith, T. G., and N. L. Buck, 1961, *Mechanical Measurements*, Addison-Wesley, Reading, Mass.

Brophy, J. J., 1977, *Basic Electronics for Scientists*, McGraw-Hill, New York.

Doebelin, E. O., 1975, *Measurement Systems*, McGraw-Hill, New York.

Holliday, C. T., and R. L. Stow, 1968, *Fiber Optics*, *S.P.I.E. Seminar Proceedings*, vol. 14.

Katz, H. W., ed., 1959, *Solid State Magnetic and Dielectric Devices*, Wiley, New York.

Lion, K. S., 1959, *Instrumentation in Scientific Research, Electrical Input Transducers*, McGraw-Hill, New York.

Lion, K. S., 1975, *Elements of Electrical and Electronic Instrumentation: An Introductory Textbook*, McGraw-Hill, New York.

Lisitsa, M. P., Berezhinskii, L. I., and M. Ya. Valakh, 1972, *Fiber Optics*, Israel Program for Scientific Translations, New York.

Millman, J., and S. Seely, 1951, *Electronics*, McGraw-Hill, New York.

Nanavati, R. P., 1975, *Semiconductor Devices*, Intext, New York.

Neubert, H. K. P., 1975, *Instrument Transducers*, Clarendon Press, Oxford.

Norton, H. N., 1969, *Handbook of Transducers for Electronic Measuring Systems*, Prentice-Hall, Englewood Cliffs, N.J.

Oliver, F. J., 1971, *Practical Instrumentation Transducers*, Hayden, New York.

Siegman, A. E., 1971, *An Introduction to Lasers and Masers*, McGraw-Hill, New York.

Welkowitz, W., and S. Deutsch, 1976, *Biomedical Instruments: Theory and Design*, Academic Press, New York.

Westinghouse Staff, 1948, *Industrial Electricity Reference Book*, Wiley, New York.

Zworykin, V. K., and E. G. Ramberg, 1949, *Photoelectricity and Its Application*, Wiley, New York.

4

***

Applications of Transducers

4.1.   INTRODUCTION

In Chapter 3, we examined various types of transducers with respect to the basic devices. In this chapter, we will examine how these devices can be applied to specific measurement applications. Initially, simple techniques are illustrated to indicate the principles involved. Toward the end of the chapter, some of the more sophisticated arrangements used in practice are presented.

4.2.   MEASUREMENT OF DISPLACEMENT

A simple technique for measuring mechanical displacement is illustrated in Fig. 4.1(a). This involves the use of a simple, rigid mechanical linkage, a resistor with a slider, a battery, and a voltmeter. The apparatus is

57

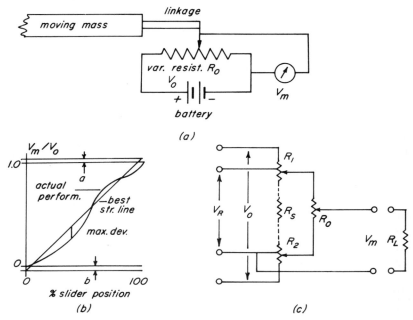

FIG. 4.1 Variable resistance transducer: (a) basic device; (b) linearity characteristic of a potentiometer compared with best-fit straight line (U.S. Army ORDP-20-137); (c) typical trimming circuit for linear potentiometer to increase linearity (U.S. Army ORDP-20-137); R_s is an optional shunting resistor.

oriented such that the direction of slider motion and the mechanical displacement to be measured are in the same direction (colinear).

$$V_m = V_0 R / R_0$$

If l_0 is the mechanical length of the resistance element R_0, and l is the distance that the wiper arm travels, we can also write

$$V_m = V_0 l / l_0$$

since $l \sim R$.

The transducer is nothing more than a voltage divider and the voltage indicated by the voltmeter is linearly dependent upon the displacement, provided that the resistor is linear. Use of a slide wire rather than a helical wound resistor would provide a higher degree of accuracy since the resistance of the slide wire should change uniformly as the slide moves, while a step change will occur with the wound resistor. To be sure, the device as presented in Fig. 4.1(a) is rather crude, but modifications of this are used in servomechanism instrumentation. If necessary, the sensitivity of the voltmeter can be increased by use of a differential voltmeter or potentiometer.

The discussion so far has assumed a linear resistance. Resistors are typically linear to some percentage (0.1–1% in this type of equipment). Figure 4.1(b) illustrates a typical nonlinearity. A trimming circuit can be used to improve linearity as shown in Fig. 4.1(c), where V_R is the reference voltage, V_0 and V_m are as above, R_1 and R_2 are center-tapped trimmer potentiometers, R_s is a shunt resistor, and R_L is the load resistor (assumed to be constant). The basic empirical relation for this circuit is

$$\text{maximum percentage error} = 15R_0/R_L$$

which maximum occurs at about two-thirds travel of the slider. If $R_L \sim 15R_0$, the maximum error is 1%, as indicated by the empirical relation. Additional circuits (as shown in Oliver, 1971) have been developed to reduce the effects of a changing value of the load (R_L).

Norton (1969) presents details of potentiometer construction, including the wiper arms, to insure maximum linearity of these devices. Continuous element (Cermet) linear resistance elements are now available in which resistance varies continuously with displacement.

The resistance strain gauge discussed previously can be used to measure displacement or strain. If we recall the expression which was developed in section 3.2.1.

$$\frac{dR/R}{\epsilon_a} = 1 + \frac{d\rho/\rho}{dl/l} + 2\mu$$

$$G = 1 + \frac{d\rho/\rho}{dl/l} + 2\mu = \text{gauge factor}$$

and if we now replace dR/R by $\Delta R/R$, then the above expression may be rewritten as

$$\frac{\Delta R/R}{\epsilon} = G$$

therefore

$$\epsilon = \frac{\Delta R}{GR}$$

Normally, the gauge factor G and the base resistance R are specified by the gauge manufacturer. The quantity ΔR may be measured with a

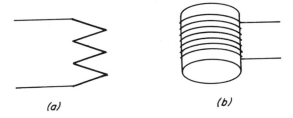

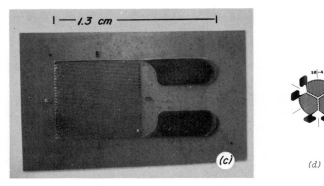

FIG. 4.2   Typical strain gauges: (a) wire gauge; (b) wire gauge wound on flattened cylindrical form; (c) metalfilm® (Automation Industries, Inc.) foil type; (d) BLH 3-element 60° planar foil rosette gauge. (Figure courtesy BLH Electronics, Inc., Waltham, MA.)

resistance bridge, so that the displacement or strain is determined directly. Typical gauge configurations are shown in Fig. 4.2.

The gauge element is attached to the object, in which strain is to be measured by means of a mounting compound which is stronger than the gauge (usually epoxy resin or a nitrocellulose cement). Orientation of the gauge or gauges is critical if one wishes to measure a particular strain quantity.

Another device frequently used for measuring displacement is the linear differential transformer (LVDT) shown schematically in Fig. 4.3. If the coils, L_1 and L_2, are identical, then for one position of the core, the output voltage is 0.

The primary winding of the transducer is usually energized by an audio signal (frequently, 1–3 kHz) and the output voltage, E, is a function of the position of the core. The coils L_1 and L_2 are bucking so that a zero output is possible

$$E = E_1 - E_2$$

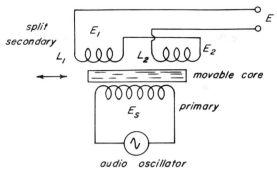

FIG. 4.3   Linear voltage differential
transformer (LVDT) schematic.

If the core is connected mechanically to a member whose displacement is to be measured, then one finds that E is linearly proportional to the core displacement, and hence the displacement of the object.

A typical voltage output–displacement curve for an LVDT is shown in Fig. 4.4. Notice that there is a phase shift of 180° at the reference position crossover. Linearity to 0.1% is attained in some commercial LVDTs. Sensitivities, measured in volts per inch (or volts/0.001 in.) vary widely. Sensitivity is frequency dependent, and manufacturers usually supply performance data for a particular frequency.

In a correctly designed LVDT, the phase angle is constant over the linear range on either side of the null position for constant frequency; the phase angle does vary as a function of frequency. If we use the network representation for an LVDT, as illustrated in Fig. 4.5, the output-to-input voltage transfer function is

$$\frac{E}{E_s} = \frac{j\omega R_L(M_2 - M_1)}{R_p(2R_s + R_L) + j\omega[L_p(2R_s + R_L) + 2R_pL_s] - \omega^2[(M_1 - M_2)^2 + 2L_pL_s]}$$

where R_L is the external load resistance, R_p is the series resistance of the primary coil, R_s is the series resistance of each secondary coil, L_p is the inductance of the primary coil, L_s is the inductance of each secondary coil, M_1 and M_2 are the indicated mutual inductances, ω is the radian frequency of operation. As a function of frequency, the primary and secondary phase angles (θ_p and θ_s) vary as follows:

$$\theta_p = \tan^{-1}(-\omega L_p/R_p)$$
$$\theta_s = 90° - \tan^{-1}(\omega L_p/R_p)$$

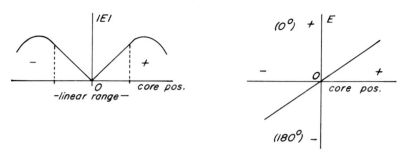

FIG. 4.4   Voltage-displacement characteristics of LVDT.

In practical applications, the null voltage can be excessive. Various circuits have been developed, as shown in Fig. 4.6, for reducing the null point residual voltage, including a servo-type circuit. This latter circuit gives maximum resolution. LVDTs differ from linear resistors in that the output voltage varies continuously rather than discretely with displacement. With a suitable null balance circuit, such as the servo system illustrated, output resolution to 0.1% is possible. Should adjustment of phase angle be required, the circuits shown in Fig. 4.7 can be used. The Rs and Cs are selected experimentally to achieve the phase angle desired.

A variable capacitance technique is applicable to the measurement of displacement as indicated in Fig. 4.8. The moving object may form one electrode of the capacitor or a special linkage and plate may be attached to the object to form the movable electrode. The limitations on this technique were discussed in detail in section 3.2.2.

The system shown in Fig. 4.8(a) is a substantial oversimplification. It is valid only at low frequencies. As frequency is increased, the dielectric losses, electrode inductance, and resistance become significant, producing the circuit representation shown in Fig. 4.8(b), where R_s is the series resistance of the leads and electrodes, L is the series inductance of the leads, C is the capacitance of the transducer, R_p is the

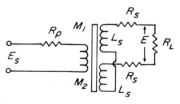

FIG. 4.5   Electrical network representation for LVDT.

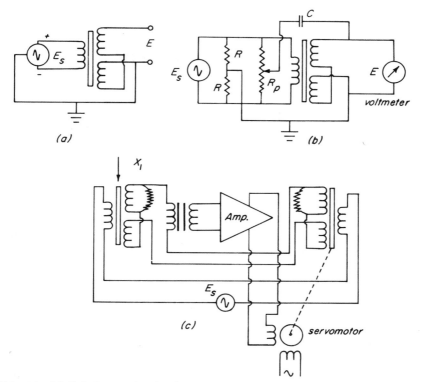

FIG. 4.6 Null balance circuits for LVDT; (a) with balanced source voltage; (b) with single-ended source voltage; (c) servo type produced by Automatic Timing & Controls, Inc. For a fixed displacement X_1 of the left slug, the servo-motor positions the right LVDT slug.

shunt resistance loss of the dielectric. If air is the dielectric, generally R_p can be ignored. If we have a double-layer capacitive transducer with a variable air gap and an insulating layer on one of the electrodes [Fig. 4.8(c)], the fractional change in capacitance is expressed by:

$$\frac{\Delta C}{C} = \frac{S\,\Delta l_1}{(l_1 + l_2)[1 - S\,\Delta l_1/(l_1 + l_2)]}$$

where

$$C = \frac{\epsilon_0 A}{l_1 + l_2/\epsilon_2} \quad \text{and} \quad S = \frac{l_1 + l_2}{l_1 + l_2/\epsilon_2}$$

Here, S is a sensitivity and nonlinearity factor that depends upon the

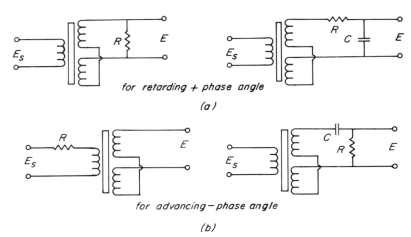

for retarding + phase angle

(a)

for advancing − phase angle

(b)

FIG. 4.7 LVDT phase-angle reducing circuits; (a) circuit to retard positive phase angle; (b) circuit to advance negative phase angle.

ratio of the respective thickness of the air gap and the dielectric coating and the dielectric constant ϵ_2 of the coating; C is simply the equivalent series capacitance of an air capacitor of spacing l_1, and a non-air capacitor of spacing l_2 and dielectric constant ϵ_2.

If in the capacitor of Fig. 4.8 we ignore R_p (air dielectric) and L [Fig. 4.8(b)], which assumes low frequency operation, we may develop some performance criteria. Let us assume that the capacitor is driven sinusoidally (a capacitance microphone driven by a sinusoidal sound wave). For this mode of operation, we do not simply measure the change in capacitance, but rather examine the change in voltage across the capacitor plates. Let us assume that the plates have a rest separation of l_0 caused by the sinusoidal sound (pressure) wave. This type of device requires a DC bias voltage as shown in Fig. 4.8(d). The transducer transfer function is given (Neubert, 1975) by

$$\frac{e_0}{f} = \frac{RE_b}{j\omega l_0 \{E_b{}^2/\omega^2 l_0{}^2 + [d + j(\omega M - K/\omega)](R + 1/j\omega C)\}}$$

where M is the mass of the movable electrode, K is the stiffness of the membrane, d is the damping, $C = \epsilon_0 A/l_0$, E_b is the DC bias voltage, R is the series resistance, and f is the applied force (from sound wave). The mechanical impedance of the transducer is

$$Z_m = d + j(\omega M - K/\omega)$$

which appears in the denominator of the dynamic transfer function shown above.

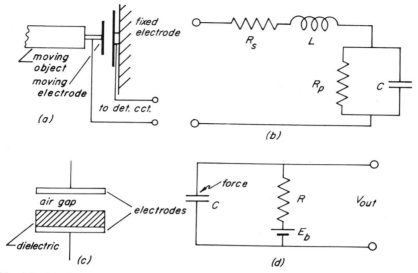

FIG. 4.8 Variable capacitance displacement transducer: (a) basic system; (b) electrical network representation; (c) unit with dielectric layer; (d) capacitance pressure sensor (capacitance microphone).

The first frequency break point in e_0/f occurs at $\omega = 1/RC$. Resonance occurs when $\omega M = K/\omega$ or $\omega^2 = K/M$. Between these two frequency limits, the device exhibits linear behavior, that is, e_0 is linearly related to the applied force f, for $\omega > 1/RC$, $R > 1/\omega C$. Thus R should be large (megohms) to extend the linear range of the device.

Construction of variable capacitors is critical. All of the component parts must be selected to insure both mechanical and thermal stability. Devices must be mechanically rigid so that dimensional tolerances are maintained. Units which have a flexible diaphragm opposing a fixed plate generally cannot be dismantled easily for cleaning or other maintenance. Thus corrosion-resistant materials should be used and the units should be appropriately sealed for protection against humidity and condensation.

4.3. FORCE MEASUREMENTS

There are three techniques commonly employed in the measurement of pressure (force/unit area) or force. These are: 1. load cell, 2. piezo-electric element, and 3. variable resistance element.

4.3.1. Load Cell

The load cell is a compressible device in which deformation must be reversible. A simple load cell consists of a metallic cylinder equipped with a strain gauge. When the cylinder is loaded, the decrease in length of the load cell (compression) is determined by the change in resistance of the strain gauge. Ultimately, the resistance change may be equated to force or pressure applied to the load cell, provided that the dimensions and elastic moduli of the cell are known.

A load cell may be constructed by using an air-filled, air-tight metal bellows in which a differential transformer is located. The body of the differential transformer is connected rigidly to the bottom of the bellows while the movable core is connected to the top. The differential transformer is adjusted such that a zero output voltage exists for zero compression of the bellows. Any compression of the bellows produces an output voltage that is calibrated as a function of the force applied to the bellows.

4.3.2. Piezoelectric Crystal

Since mechanical stressing of a piezoelectric crystal generates a voltage, one may use such a crystal as a load cell. The upper limit to pressures that can be measured in this manner is relatively low as most piezoelectric crystals fracture rather easily.

4.3.3. Variable Resistance Element

The resistance of many metallic conductors changes linearly with pressure over small ranges of pressure variations according to the relation

$$R = R_0(1 + \gamma\,\Delta p)$$

where R_0 is the resistance (at 1 atmosphere) in ohms, γ is the pressure coefficient of resistance in ohms/lb/in²., Δp is the pressure change in lb/in.². A type of load cell may be constructed using this principle, as shown in Fig. 4.9. It consists of a gas-filled cylinder and piston which contains a resistance element. The change in resistance with pressure may be detected by a Wheatstone bridge.

Semiconductor devices may also serve as strain gauge elements. The conductivity σ of a semiconductor is given by

$$\sigma = qn_i\mu'$$

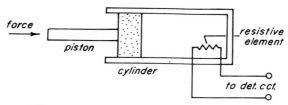

FIG. 4.9   Resistance pressure transducer.

and the resistivity by

$$\rho = (qn_i\mu')^{-1} \qquad \rho_0 = \text{initial value of } \rho$$

where q is the electric charge (charge of an electron), n_i is the number of intrinsic charge carriers (holes or electrons), and μ' is the average mobility of the charge carriers. When an external mechanical stress is applied to a semiconductor, internal changes in the crystallographic structure occur with related changes in both n_i and μ'. In the case of simple tension or compression along the stress axis, the change in resistivity $\Delta\rho$ is given by

$$\Delta\rho = \rho_0\beta_l\sigma'$$

where β_l is the longitudinal piezoresistive coefficient, and σ' is the mechanical stress. The change in electrical resistance to applied mechanical stress is called the piezoresistive effect.

The gauge factor for these devices is given (Oliver, 1971) by

$$G = \Delta R/R\epsilon = 1 + 2\mu + \beta_l Y$$

where R is the initial resistance of the gauge element, ϵ is the normal strain, μ is Poisson's ratio, and Y is Young's modulus. When silicon is used as the gauge material at temperatures below 1000°F, it provides an almost perfectly elastic material that is nearly free from nonlinear effects such as drift, creep, and hysteresis. Strain gauges are frequently used in load cells. For this application, the gauges are often fabricated in either ring or disc configurations. The appropriate strain-gauge bridge circuit is shown in Fig. 4.10(a). Normally load cells are used when very high pressures or forces are to be measured. Usually these devices would not be used in biological or medical applications, but they do find use in some environmental applications such as testing the load-bearing capacity of soils, etc.

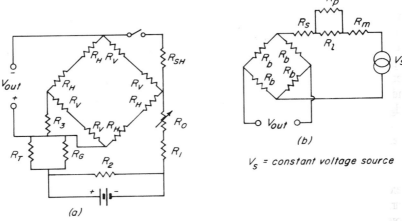

FIG. 4.10   Strain gauge transducers: (a) load cell strain gauge configuration; (b) linearization of strain gauge bridge using semiconductor element R_l. The subscripts b, H, and V indicate, respectively, bridge, horizontal, and vertical.

Strain gauge transducers exhibit some degree of nonlinearity, especially the disc and ring types used in load cells. Regardless of the strain gauge application, linearization can be achieved, in part, through the use of the circuit shown in Fig. 4.10(b). A silicon semiconductor strain-gauge element R_l is mounted on the specimen (or transducer element, depending upon application). If $R_l < R_m + R_s + R_b$, then R_l is selected according to the relation

$$R_l = \frac{\%D(R_b + R_s + R_m)}{G\epsilon}$$

where $\%D$ is the $\%$ deviation from linearity, R_b is the resistance of the strain gauge bridge, R_s is the range setting resistance, R_m is the correction resistance for elastic modulus, G is the gauge factor for the semiconductor resistor, and ϵ is the cell strain level. The semiconductor strain gauge R_l is affected to some extent by temperature; however, adjusting R_m after R_l is in the circuit will correct for this provided that temperature does not change during measurement.

4.3.4.   Further Discussion of Piezoresistive Devices

A piezoresistive element is analogous to a large output signal strain gauge. Semiconductor materials are now generally used varying from p-doped bulk silicon to semiconductors with surface-diffused impurity regions. Transducers designed for medical applications normally exhibit cantilever configurations. The active elements are thin strips of semiconductor material firmly clamped at one end. Mechanical displacement of the opposite end by directly applied force, or indirectly by pressure against an effective diaphragm, produces changes in the electrical conductivity of the material. Detection of this change in conductivity (or electrical resistance) is by conventional Wheatstone bridge techniques.

A simple mathematical treatment for these devices is based upon the electromagnetic field theory formulation of Ohm's Law:

$$\mathbf{J} = \sigma \mathbf{E}$$

where $\mathbf{J}$ is the current density (Am^{-2}) in the material and $\mathbf{E}$ is the electric field (Vm^{-1}) resulting from the applied potential gradient across the device. The electrical conductivity σ (Siemens) is device-orientation sensitive so that we have the defining relations (relative to cartesian coordinates)

$$\begin{bmatrix} J_x \\ J_y \\ J_z \end{bmatrix} = \begin{bmatrix} \sigma_{xx} & \sigma_{xy} & \sigma_{xz} \\ \sigma_{yx} & \sigma_{yy} & \sigma_{yz} \\ \sigma_{zx} & \sigma_{zy} & \sigma_{zz} \end{bmatrix} \begin{bmatrix} E_x \\ E_y \\ E_z \end{bmatrix}$$

whereas a normal linear isotropic (Ohm's Law) resistive device would be defined by

$$[\sigma] = \begin{bmatrix} \sigma & 0 & 0 \\ 0 & \sigma & 0 \\ 0 & 0 & \sigma \end{bmatrix}$$

From the tensor nature of σ in piezoresistive elements, we see that they respond to linear, rotational, and shear stresses. This can cause some calibration problems.

Typical dimensions for cantilever configurations are of the order of 1 cm long by 0.5 cm wide by a few tenths of a millimeter thick. Mechanical resonant frequencies are usually above 1 kHz. Temperature instability may cause calibration difficulties.

4.4. TEMPERATURE TRANSDUCERS

A simple device for measuring temperature is the resistance thermometer. The physical principle involved is the change in resistance of a metallic conductor with temperature change. A resistance thermometer consists of a coil of wire wound on either a flat or cylindrical form. The exact construction of the transducer depends upon the temperature range to be measured. For biological applications, where temperatures are on the order of 30–40°C, a suitable resistance thermometer can be constructed by winding approximately a meter of number 40 Formvar-coated soft-copper wire on a fine glass form. The base resistance is on the order of 3 Ω. Such a transducer may be connected into a Wheatstone bridge in which the detector arm is a strip chart recorder. In this manner, a continuous temperature record may be obtained.

For high temperature work (up to 1000°C), platinum resistance thermometers are employed. This transducer consists of a flat ceramic form about which bare platinum wire is wound. Care must be taken to assure that adjacent turns of wire do not touch, and the form should be moisture free.

The Seebeck effect may be utilized to form a thermocouple. In this case a bimetallic junction is formed by welding two dissimilar wires together. A voltage is produced (thermal emf) in accordance with the temperature of the junction. Typical junctions are shown in Fig. 4.11. Common thermocouple materials are indicated in the table below.

Thermocouple Materials

Material 1	Material 2
Copper	Constantan
Chromel	Constantan
Iron	Constantan
Chromel	Alumel
Platinum	Rhodium

As the voltage output of a thermocouple is rather small (on the order of 1–20 mV), a potentiometer is usually used as the voltage measuring device (see section 6.5). A millivolt pyrometer may also be used.

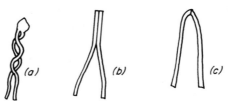

FIG. 4.11   Types of thermocouple junctions: (a) leads twisted and welded at end; (b) leads flattened and welded; (c) leads butt-welded.

4.5.   FLUID FLOW MEASUREMENTS

There are a number of techniques possible for the measurement of fluid flow. Two rather unsophisticated ones are presented here. The first is an acoustic method and is illustrated in Fig. 4.12.

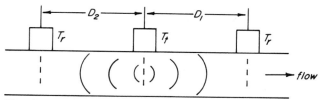

FIG. 4.12   Simple transit time ultrasonic flowmeter: T_r is the receiving transducer; T_t is the transmitting transducer; D_1 and D_2 are the separations between transducers.

Three piezoelectric transducers are attached to the flow pipe. The center one is energized by an electrical signal at about 1 MHz. Actually pulsed rf is used. The transducer vibrates at this frequency and transmits a shock wave to the fluid. The shock wave is detected by the two receiving transducers and converted into an electrical signal. These signals are then fed to a timing circuit that indicates the time of travel of the acoustic pulse upstream and downstream. The fluid velocity is given by

$$u_f = \frac{u_0(T_2 - T_1)}{(T_1 + T_2)}$$

where u_0 is the velocity of ultrasound in still fluid (1500 m/s), T_1 is the transit time for a downstream pulse, and T_2 is the transit time for an upstream pulse.

Another device, the electromagnetic flowmeter, utilizes Faraday's Law. The configuration of the transducer is shown in Fig. 4.13. A conducting fluid, such as blood, is required for operation of the device. Motion of the conducting fluid through the magnetic field generates an emf that is detected across the probe electrodes. If a permanent magnet or a DC energized electromagnet is used, a direct voltage is produced that is directly proportional to the fluid velocity. An AC electromagnet is preferable in many cases. If AC is used, however, a transformer emf will be observed, i.e., a voltage can be measured across the electrodes when the fluid is not in motion. Special circuitry is required to cancel the effect of the transformer emf. The generated emf is given by the expression.

$$V = kBdu_f \times 10^{-8}$$

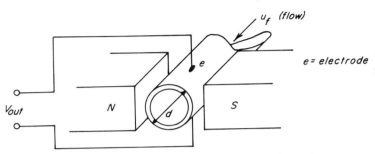

FIG. 4.13   Electromagnetic flowmeter configuration.

where V is the generated voltage in volts, k is a factor involving the electrical parameters of the fluid, d is the pipe diameter in cm, u_f is the fluid velocity in cm/sec, and B is the field strength in gauss.

4.6.   MEASUREMENT OF ACCELERATION, SHOCK, AND VIBRATION

One simple transducer may be used to provide an electric output for the mechanical inputs noted above. Shock and vibration can be described in terms of displacement, velocity, or acceleration. Acceleration is a convenient quantity to use since it is intimately related to force through Newton's Second Law of Motion,

$$\mathbf{F} = d\mathbf{p}/dt$$

where $\mathbf{p}$ is momentum. Frequently, one wishes to know both force and acceleration in cases where destructive forces are involved.

The simple piezoelectric transducer may be used with excellent results as an accelerometer. The crystal is mounted in a rigid housing which in turn is positively attached to the object whose acceleration is to be measured. The mechanical stress applied to the piezoelectric crystal by shock, vibration, or acceleration is converted into an electrical output. Typical response characteristics for piezoelectric transducers are:

Frequency response	2–12,000 Hz
Dynamic range	0.001–10,000 g
Internal resistance	20,000 MΩ

Because of the high internal resistance of the piezoelectric element, high-input-impedance amplifiers or followers are usually required to couple the transducer output to the electrical detection system. Operational amplifiers connected in the follower mode are highly satisfactory. Calibration of piezoelectric accelerometers is critical.

It should be noted that there are numerous transducers for detecting acceleration. Only one is included here because of our need to restrict coverage to electrical output devices.

4.7.   ELECTROMAGNETIC DOSIMETER

As was pointed out at the beginning of this article, there are numerous forms of transducers. Those which produce electrical outputs for various types of inputs represent a small segment of the overall population. For this example, we will deviate from the constraint that the transducer output must be electrical in order to indicate the variety which exists in transducers.

Anyone who has had any connection with X-rays or radioactive material is familiar with the film badge detector worn by people working with radiation. The electromagnetic energy (ionizing radiation) impinging upon the film causes cloudiness in the film upon development. If one knows the length of time of exposure of the film to the radiation, then a direct measure of radiation intensity as a function of film opacity is possible for fixed development time. Thus we have a transducer, the film, which converts an electromagnetic input to a permanent optical output.

In the radar range of electromagnetic energy (500 MHz to 30 GHz) health hazard problems also exist. The vector here is not ionizing radiation, but rather thermal damage to biological tissue as a result of the absorption of the em energy. It has been ascertained that a power flux of 1 mW/cm² is the maximum safe dose for total body irradiation. The question now arises as to how this power can be measured. Of course a radio-frequency power meter could be used, but these are unwieldy and extremely frequency sensitive. A simple device, similiar to the film badge, which would give a positive indication is preferable. Such a transducer is possible, as the following model system will demonstrate. A small container is filled with a liquid which simulates the average electrical properties of biological material (the human body). Certain organic dyes have the property of changing color rapidly as a function of temperature. If one of these dyes is added to the body-simulant fluid, then by the color of the dye, one has a gross measure of the fluid temperature, and hence, the absorbed power. Here we have an electromagnetic input, which produces a thermal change, which causes a chemical change, which produces an optical output.

4.8.   HALL-EFFECT DEVICES

Semiconductors such as InSb and InAs exhibit the property of magnetoresistance, that is, their electrical conductivity changes in response to an applied magnetic field. Let us examine a semiconductor specimen with the configuration shown in Fig. 4.14. A current I is established in the specimen by the battery voltage V. A magnetic field **H** is applied at right angles to the current. We assume that the specimen is not magnetostrictive so that its physical dimensions do not change under the influence of the applied magnetic field. If electrodes are placed on the top and bottom of the specimen (points a and b), a voltage, the Hall voltage, can be measured. The magnetic field causes charge carriers to be deflected along an axis mutually perpendicular to the directions of **H** and I thus establishing an electric field E_{ab}, such that

$$E_{ab} = R_H HJ \text{ V/m}^2$$

$$J = \frac{I}{A}$$

where R_H is the Hall constant (A-Ω/m), H is the applied magnetic field intensity (A/m), I is the applied current (A), J is the current density

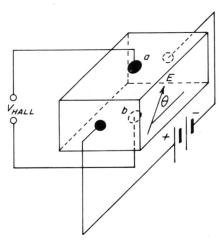

FIG. 4.14   Demonstration of the Hall effect. **H** is applied from left-to-right; battery is V volts.

(A/m²), A is the cross-sectional area of the specimen (m²), and the general relation is

$$\mathbf{E} + R_H \mathbf{J} \times \mathbf{H} = \rho_H \mathbf{J}$$

where ρ_H is the specimen resistivity as a function of H (Ω-m). In the example we have chosen, $\mathbf{E}$ and $\mathbf{H}$ are at right angles. We can now derive a relation for the angle θ between $\mathbf{E}$ and $\mathbf{J}$. We take the cross product of both sides of the general relation and combine as follows:

$$\begin{aligned} \rho_H \mathbf{E} \times \mathbf{J} &= \mathbf{E} \times \mathbf{E} + R_H \mathbf{E} \times (\mathbf{J} \times \mathbf{H}) \\ &= R_H [\mathbf{J}(\mathbf{E} \cdot \mathbf{H}) - \mathbf{H}(\mathbf{E} \cdot \mathbf{J})] \\ &= -R_H (\mathbf{E} \cdot \mathbf{J}) \mathbf{H} \quad \text{since } \mathbf{E} \cdot \mathbf{H} = 0 \end{aligned}$$

Expanding

$$\rho_H EJ \sin \theta = -R_H EJH \cos \theta$$

Therefore

$$\tan \theta = -HR_H / \rho_H$$

The electrical field at the specimen's center is oriented at an angle θ relative to the current. By using geometries other than a rectilinear bar, we can increase the apparent change in resistivity and V_{Hall}.

This effect can be utilized to build a transducer to measure magnetic field intensity since $V_{\text{Hall}} = kH$, where k is a proportionality constant.

4.9. MAGNETOSTRICTIVE DEVICES

Piezoelectric devices, as discussed in section 3.3, are electrostrictive devices in which an applied electric field produces changes in the physical dimensions of the device. A similar process occurs when a magnetic material is subjected to an applied magnetic field. Figure 4.15 illustrates one transducer application of this phenomenon. A number of thin iron laminations are bound together and subjected to the magnetic field. When the current is applied, the length of the laminations changes. If a sinusoidal current is applied, then the elongation of the laminations is sinusoidal. This principle is used in fabricating the driver transducers for magnetic sonicators used to destroy cellular matter. The vibratory motion of the laminations is translated to the liquid which contains the cellular material, which in turn shakes the cells apart, destroying the cell membranes. The mathematical treatment of these devices is quite complex and is omitted here.

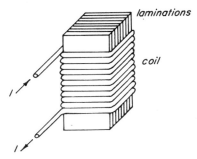

FIG. 4.15 A simple magnetostrictive device using ferromagnetic laminations and an energizing coil carrying a current I.

4.10. EXAMPLE OF A COMPOUND TRANSDUCER

One example of a compound transducer is the instrumentation employed to detect the displacement of a ballistocardiograph table. The table is essentially a hammock that is suspended from a rigid support by wires attached to its four corners. The mechanical impulses produced by the heartbeats of a patient lying upon the table cause motion of the assembly. Indirect means must be used to detect the displacement of the table, as any direct mechanical connection would introduce severe damping of the motion. A successful technique for detecting the table motion is illustrated in Fig. 4.16. An opaque vane, which interrupts a collimated light beam, is attached to the table. In this manner the mechanical motion of the table is converted into a varying light intensity. The varying-intensity light beam is focused upon the photosensitive surface of a photoelectric detector, which then produces an electrical output.

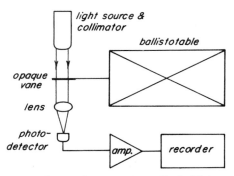

FIG. 4.16 A compound transducer system: the ballistocardiograph table.

Thus we have the following sequence of conversions: mechanical input to optical output, optical input to electrical output, and electrical input to mechanical output to motion of recorder pen.

In this chapter, we have examined some simple transducers and applications. In Chapter 5, we will look more deeply into transducers and transducer systems that are used in medical and environmental applications.

4.11. REFERENCES

Doebelin, E. O., 1975, *Measurement Systems*, McGraw-Hill, New York.
Herceg, E. E., 1972, *Handbook of Measurement and Control*, Schaevitz Engineering, Pennsauken, N.J.
Lion, K. S., 1975, *Elements of Electrical and Electronic Instrumentation.* McGraw-Hill, New York.
Neubert, H. K. P., 1975, *Instrument Transducers*, Clarendon Press, Oxford.
Norton, H. N., 1969, *Handbook of Transducers for Electronic Measuring Systems*, Prentice-Hall, Englewood Cliffs, N.J.
Oliver, F. J., 1971, *Practical Instrumentation Transducers*, Hayden, New York.

Specific Transducers for Medical and Environmental Applications

5.1. INTRODUCTION

In Chapters 3 and 4, we examined the basic principles of operation of transducers. Now we will see how the components are arranged to produce practical devices with clinical and environmental applications.

5.2. CLINICAL PRESSURE MEASUREMENTS

The normal method for measuring blood pressure is the mercury manometer sphygmomanometer. Because of the pneumatic cuff requirement of this device, it is impractical for continuous blood pressure monitoring, as might be required during certain surgical procedures and in intensive care units. Several transducer techniques may be applied

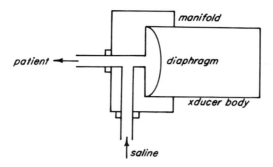

FIG. 5.1   Schematic of basic blood pressure transducer.

for continuous monitoring of arterial pressure and venous mean pressure. These are direct techniques which involve a fluid connection into a blood vessel, as opposed to the indirect or noninvasive sphygmomanometer.

Transducers for blood pressure measurements have several common features as shown in Fig. 5.1. There is a pressure diaphragm, a surrounding manifold with connections for a catheter to the patient, and connections to a saline flush bottle. The transducer body may house one of several possible transducer systems.

One transducer system involves the use of four resistive strain gauges connected as a Wheatstone bridge. The active gauge is attached to the pressure diaphragm. The other three resistive elements are placed in the transducer body so that all four resistors are at the same temperature. This avoids differential changes in resistance as a function of temperature. A hypodermic needle is connected to a catheter tube which, in turn, is connected to the manifold of the transducer. Another tube is connected from the manifold to a bottle of sterile saline solution. Saline is allowed to flow through the system until it passes out of the tip of the needle, which is then inserted into the subject's blood vessel. The flow from the saline bottle is then shut off at the manifold, leaving a saline column between the patient and the pressure diaphragm. Changes in blood pressure are transmitted through the saline column to produce changes in the diaphragm's position. This in turn flexes the attached strain gauge to produce a change in resistance as a function of blood pressure. The Wheatstone bridge is initially balanced, so that resistive changes of the active gauge produce an imbalance voltage across the detector arm of the bridge. Normally the bridge is energized with a 1–2 kHz sinusoidal voltage. The pressure changes then produce an output voltage whose amplitude varies according to the pressure

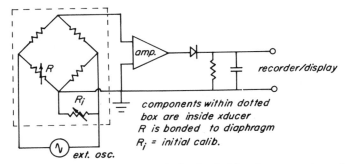

FIG. 5.2   Electrical circuit schematic for basic blood pressure transducer.

changes. This voltage is amplified and then rectified to produce a signal waveform which is directly proportional to pressure. This waveform may be displayed on an oscilloscope or a strip-chart recorder.

Figure 5.2 illustrates a possible circuit schematic for this system, and Fig. 5.3 a typical output signal. The transducer has relatively high mechanical compliance so that it responds rapidly. The saline solution is considered incompressible, so that the pressure changes are transmitted faithfully. There is a slight differential pressure loss because of fluid friction associated with the tube walls.

An alternative transducer system utilizes the basic features described above, but the Wheatstone bridge is replaced by a differential transformer as described in section 4.2. A circuit is shown in Fig. 5.4. Because of the mechanical inertia of the differential transformer core, this system has a lower compliance than the strain gauge system, and is thus less responsive. A typical device is shown in Fig. 5.5.

A second alternative uses a variable capacitance technique. The pressure diaphragm forms the movable plate of the capacitor. A fixed plate is mounted in close proximity. The capacitor forms part of the tuned circuit of either a Colpitts oscillator or autodyne detector. The electronic circuitry for the oscillator is contained within the transducer.

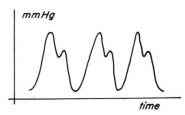

FIG. 5.3   Idealized blood pressure record from blood pressure transducer.

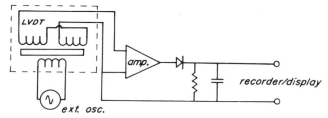

FIG. 5.4   Electrical circuit schematic for LVDT blood pressure transducer.

The variable capacitance changes the oscillator frequency, which is then detected by a linear discriminator circuit (for the Colpitts oscillator) or directly converted in the autodyne circuit. The voltage is then amplified and displayed. A capacitance bridge might also be used, but it is not as sensitive a system. This system has a high mechanical compliance, but suffers from temperature effects. Changes in temperature cause the base capacitance value to change, thus offsetting the oscillator frequency. A basic system is shown in Fig. 5.6.

Some of the dimensional stability problems associated with variable capacitance devices have been overcome in a differential capacitance transducer design developed by Trans-Sonics, Inc. of Burlington, Mass.

FIG. 5.5   Typical LVDT-type blood pressure transducer (courtesy Sanborn Division, Hewlett-Packard Co., Waltham, MA).

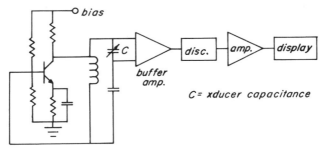

FIG. 5.6 Electrical schematic for blood pressure transducer system using a variable capacitance transducer and Colpitts oscillator.

This device is illustrated in Fig. 5.7. The fixed plates of the capacitor are formed from glass discs in which spherical depressions have been ground. These are then gold plated to form the stator plates. The movable electrode is a stainless steel diaphragm stretched between the stators. For blood pressure measurements, the steel diaphragm could be replaced by a metallized plastic film for higher compliance. In normal operation, the right orifice of the transducer would be sealed, so that the blood pressure works against a fixed air cylinder as in the scheme represented in Fig. 5.1.

Generally a bridge detector, as shown in Fig. 5.8, is used with this system, which is sold commercially as the Equibar differential pressure transducer. The output from the bridge is connected to a follower circuit (Chapter 7) and the pressure signal is detected using either a conventional demodulator for normal measurements, or a phase-sensitive demodulator and filter for differential measurements (Chapter 8).

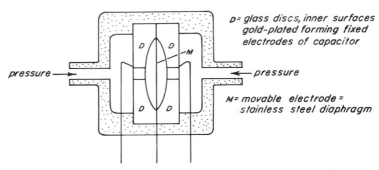

FIG. 5.7 Trans-Sonics, Inc. differential capacitance transducer.

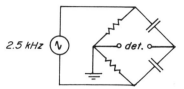

FIG. 5.8   Bridge detector circuit for use with differential capacitance transducer.

5.3.   BLOOD CELL COUNTER

The principle of varying a resistance can be used to advantage for counting blood cells. A blood cell, because of its membrane, is basically a poor conductor at low electrical frequencies, despite the fact that the internal cytoplasm is a good electrical conductor. Blood serum is an electrolyte and hence a good conductor. Thus a small volume of plasma that contains blood cells will exhibit a lower electrical conductivity (higher resistance) than the same volume of plasma alone. Based upon this property, a counting device can be constructed as shown schematically in Fig. 5.9. Normally a whole blood sample is diluted using an isotonic electrolyte diluent. A vacuum pump is used to draw a calibrated volume of solution through a small orifice, usually on the order of 100μm. The orifice is effectively a variable resistance transducer, in which resistance varies as a function of the number of cells passing through the orifice. The resistance change causes a current change in the external electrical circuit, which in turn produces a varying voltage across the resistor R. The varying voltage will be of pulsatile form, with the pulse

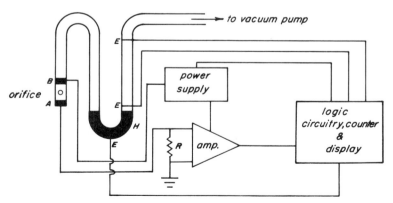

FIG. 5.9   Schematic for blood cell counting system.

amplitude proportional to the number of cells passing through the orifice. The voltage across R is then electronically processed by threshold and pulse amplitude discriminator circuits, and the resultant output directed to an electronic counter and display. The threshold is manually adjustable so that either red or white blood cells (RBCs or WBCs) can be counted. The amplitude discriminator in effect measures the number of cells as a function of pulse amplitude.

The calibrated sample volume is achieved by using the vacuum to raise an electrically conducting fluid column between two electrodes placed at the ends of a calibrated glass tube. When the top of the fluid column contacts the lower electrode, this acts as a switch and the electronic counter is activated. When the top of the fluid column contacts the upper electrode, the counter is turned off. In this manner, counts per unit volume are obtained, since the amount of fluid between the two electrodes is equal to the amount of sample drawn through the orifice. Mercury is frequently used as the conducting material.

The connections to the orifice "variable resistance" are achieved by placing an external and internal metal band on the orifice tube. The entire resistance is then composed of the orifice and the resistance of a small amount of fluid on either side of the orifice.

5.4.   HEMOGLOBINOMETER (HEMATOCRIT DETERMINATION)

In this device, an optical principle is used. Hemoglobin strongly absorbs light in the 540 nm range when prepared in a cyanmet lysing solution for RBCs. A known volume of blood is diluted with the reagent to lyse the blood cells and oxidize the hemoglobin to cyanmethemoglobin. The light absorbance of the resulting solution occurs at 540 nm and is proportional to hemoglobin content in the sample.

The hemoglobinometer is constructed as a special purpose spectrophotometer using a light emitting diode (LED) as the light source and a light detecting diode (LDD) as the optical detector. Commercially available LEDs and LDDs are available in the 560 nm range, which is sufficiently close to 540 nm. A light-tight chamber holds the LED and LDD on opposing faces with the sample cuvette interposed between them. The remaining electronic circuitry, shown in Figure 5.10, consists of an amplifier to increase the signal from the LDD, and a meter. In use, the cuvette is first filled with the lysing solution to obtain a calibration reference, then the sample is measured. A typical calibration curve, obtained using a cyanmethemoglobin reference standard, is

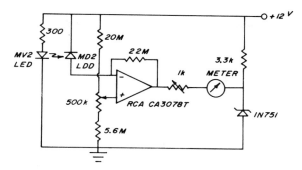

FIG. 5.10   Electrical schematic for hemoglobinometer.

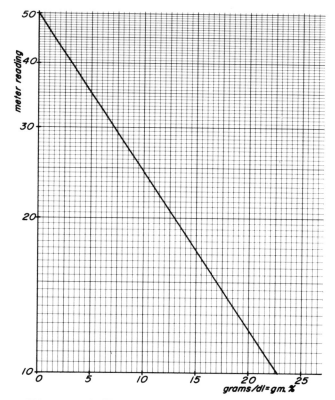

FIG. 5.11   Calibration curve for hemoglobinometer.

shown in Fig. 5.11. Hematocrit for human blood is obtained by multiplying the calibration curve reading by three.

5.5.   OXIMETRY—A COLORIMETRIC TRANSDUCER

Oxygen saturation in a highly vasculated area of tissue can be ascertained by measuring the "redness" of the tissue. A relatively simple transducer system can be constructed (Geddes and Baker, 1975), as outlined in Fig. 5.12, by transilluminating the ear lobe. A broad-spectrum light source is used on one side of the ear lobe. Two photosensitive pickups are placed on the opposite side; one sensitive in the infrared range (ca. 800 nm) and the other in the visible red range (ca. 640 nm). The output from the IR channel yields information on blood and tissue volume in the optical path; it does not respond to oxygen level. The visible red channel gives information on oxygenation of the blood. As oxygen increases, the blood becomes more red and more light is absorbed over the optical path. The device must be calibrated against chemically analyzed blood samples. Practical self-contained units which clip onto the ear have been developed.

5.6.   PIEZOELECTRIC TRANSDUCERS

Piezoelectric devices find various applications in medical diagnostic work. Two are discussed here and others will be described in subsequent chapters. The first application uses a piezoelectric crystal in the accelerometer mode to detect the radial pulse for continuous monitoring of a patient. It has several advantages over ECG monitoring since no electrode paste, and only one contact with the subject, is required. The transducer element is mounted on a band similar to a wristwatch band. This is placed about the wrist with the piezoelectric element over that area of the wrist where the radial pulse is manually detected. The output from the transducer is amplified and connected to a monitor, either to display the signal or to count pulses per unit time to determine pulse rate. Figure 5.13 diagrams a basic system.

Another application of these devices is in the monitoring of heart sounds, either in an electronic stethoscope or for the recording of a phonocardiogram. A sensitive piezoelectric crystal is mounted in a sound shield similar to the bell of a conventional stethoscope (a larger shield is used to record phonocardiograms). The unit is placed on the body surface in the same manner as a conventional instrument. The

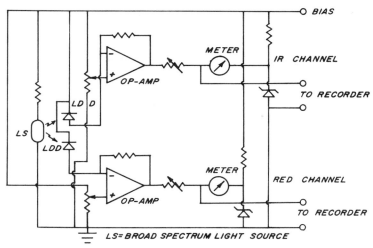

FIG. 5.12 Electrical schematic for ear clip oximeter.

signal from the crystal element is electronically amplified and connected to crystal (piezoelectric) earphones. For phonocardiography, the amplified output is directed to a paper chart recorder or an oscilloscope. Some filtering of the heart sounds may be employed in order to visualize certain acoustical patterns.

In certain applications, a more sensitive pickup can be made by using a variable capacitance transducer, commonly known as a capacitance microphone (discussed previously). This is composed of a fixed plate and a very high compliance diaphragm. Either very thin metallic foil is employed for the diaphragm, or mylar film on which metal has been deposited. Detection of the sonic energy is based on an autodyne oscillator technique.

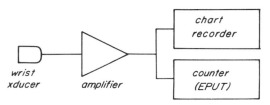

FIG. 5.13 Radial pulse counter using piezoelectric transducer (see Geddes and Baker, 1975).

5.7.   CLINICAL THERMOMETER

For convenience, electronic thermometers have replaced conventional mercury thermometers in many hospital applications. The device consists of a thermistor resistance probe connected into a bridge circuit as shown in Fig. 5.14. A thermistor is used that exhibits a reasonably linear characteristic over the clinical range of body temperatures. An initial calibration control is provided to balance the bridge for a designated reference temperature of the probe. The instrument scale is calibrated to reflect the characteristic of the probe. An alternative approach is to use a logarithmic amplifier whose characteristics inversely match those of the probe. In this manner, the thermistor response is linearized.

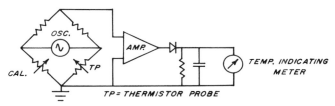

FIG. 5.14   Resistance thermometer (thermistor) bridge for temperature measurements.

5.8.   RESPIRATORY MONITORING

Monitoring of the respiration rate presents some interesting problems. Basically one must decide what is the appropriate quantity to use as an excitation for a transducer. The expansion and contraction of the chest is one possibility. Another possibility is to use the breath itself. Transducer systems can be constructed along both of these bases.

　　If chest expansion, as related to respiration, is to be monitored, then several possibilities exist. All are implemented by using a resilient band placed around the chest. An early technique used a tube, containing mercury, which was connected to a Wheatstone bridge as the "unknown" arm. Normal respiration produced changes in the length and cross-section of the tube, thus changing the electrical resistance of the mercury column. This resistance change produced a voltage across the detector arm of the bridge which varied with respiration.

　　A modern approach to the mercury column uses a conducting silastic rubber band that replaces the mercury. The dimensions of the band change with respiration, producing a resistance change in direct proportion to breathing. The detection system is the same as for the mercury column.

Another approach that has been tried is a pneumatic system. An air-filled tube is used in conjunction with a pneumatic switch. Pressure changes in the tube activate the switch which in turn is connected to a voltage source and events per unit time counter. The first two systems permit monitoring the depth of respiration as well as rate since the resistance change is proportional to the extent of chest expansion. The pneumatic system permits rate monitoring only.

A fourth monitoring system uses a thermistor placed next to a nostril. The change in air flow produces a temperature change in the thermistor sensor, which in turn produces a resistance change that is detected as in the clinical thermometer. The outputs from all four of the systems mentioned can be directed to an appropriate pulse generating circuit, such as a Schmitt trigger, for counting by an EPUT (events per unit time) meter. In this manner, respiration rate is obtained.

5.9.   ENVIRONMENTAL MONITORING

Equipment used in environmental monitoring is just now beginning to reach the level of sophistication that has been the standard for instrumentation used in other areas. Many instruments are still rather basic, and the data obtained from them more qualitative than quantitative. Some interesting technical problems are encountered in their design and several of these are explored in the remaining sections of this chapter. Many aspects of water quality require pH and chemical analyses. Some of these tests are carried out using ion-specific electrodes, which are treated in a subsequent chapter. For the most part, we will examine small, portable instruments that can be used in the field rather than in the laboratory only.

5.10.   EXPLOSIVE LIMIT DETECTOR

An ever-present problem in the mining industry is the accumulation of explosive gases in enclosed areas. Although gas analyzers are available, they are essentially laboratory rather than field instruments. Before we describe an instrument that detects explosive mixtures, it is necessary to characterize some basic properties of such mixtures. Not all combinations of combustible materials dispersed in air will explode. There is a concentration threshold called the lower explosive limit (LEL) below which no explosion occurs when ignition is attempted. Above the LEL, there is a range of concentrations in which explosion will occur. This includes all concentrations for which a flash will occur or a flame will travel when the mixture is ignited. Mixtures can reach a point when they are too rich to ignite. The concentration at which this occurs is called the upper explosive limit (UEL). We express explosive limits as percent by volume of vapor in air. Tables are available that define the LEL and UEL values for mine gases and for industrial gases and vapors.

The portable device used to evaluate explosive mixtures is called an explosimeter. In principle, it permits evaluation of combustibility in a controlled environment. The device is shown schematically in Fig. 5.15.

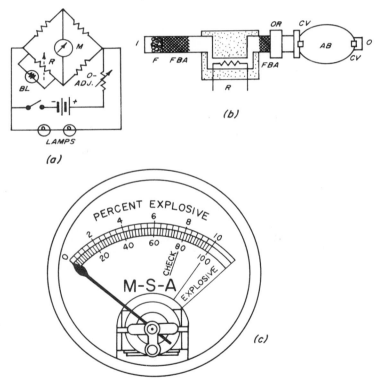

FIG. 5.15   Explosimeter (Mine Safety Appliances Co.). (a) Electrical schematic: BL = constant current ballast lamp; R = hot filament; M = indicating meter. (b) Mechanical design: AB = aspirator bulb; CV = check valve; F = filter; FBA = flashback arrestor; I = inlet; O = outlet; OR = orifice; R = hot filament. (c) Meter scale.

In operation, an amount of the air–gas mixture is introduced by an aspirator bulb into a closed chamber. Within this chamber is a white-hot filament that forms one arm of a Wheatstone bridge. With the chamber containing air only, the bridge is balanced. When the gas–air mixture hits the white-hot filament, it will burn if the concentration is between the LEL and UEL. Additional heat is thus released and the electrical resistance of the filament changes, producing an offset voltage in the detector arm of the bridge. Normally the meters in the bridge detector arm, as shown in Fig. 5.15, are calibrated in % LEL. In many cases the LEL, expressed in ppm, is well above the toxic limit for the material.

In using instruments of this sort, various precautions must be taken with certain gases, methane being one of them. Generally speaking, the manufacturer will provide sufficient instructions for the instrument's use; our intent here has been to present the basic technique involved.

5.11. MEASUREMENT OF STACK-PLUME EFFLUENT

As a spin-off from the space program, NASA has developed a number of instrumentation systems that find applications in the medical and environmental areas. Several of these are discussed in the sections of this chapter which follow.

Industrial installations are now required to monitor stack effluents for noxious and toxic emissions. A basic requirement is a small and portable field instrument. Under laboratory conditions, mass spectrometers and various types of gas chromatographs can be used, but these are not portable field instruments. The Langley Research Center of NASA has developed a portable solar radiometer (reference designation TN D-8182[N76-26718]) for determining the amount of NO_2 and SO_2 in stack emissions. This instrument is shown in Fig. 5.16. The measurement can be made from ground level while pointing the device at the plume. The criterion for use is that the sun be accessible to view and that the sky adjacent to the sun be completely cloud free.

The operation of this device is based upon a differential absorption measurement of the light reaching the radiometer. The solar radiation passing through the plume is measured, using four wavelength-selective filters, relative to the radiation from the sun as a background or reference source. Two sets of optics are contained in the device; one is a simple bore sight used to point the radiometer toward the sun

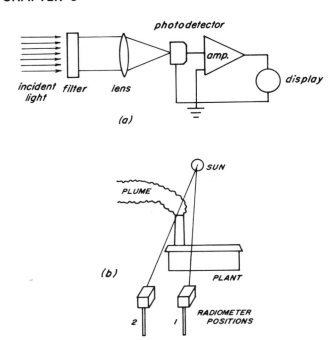

FIG. 5.16   Portable radiometer for measuring NO_2 and SO_2 in stack plumes (redrawn from *NASA Tech. Briefs*, Winter, 1976).

(for background or reference measurement) or toward the plume. The other optical path is the radiometer channel.

Four filters are used in the radiometer channel. The background, with the unit pointed at the sun, is determined and recorded as I_0 for four wavelengths: 310 nm (UV), 400 nm (visible), 600 nm (visible), and 800 nm (IR). A second measurement, at these four wavelengths, is made with an optical path through the plume using the sun as the background, as shown in Fig. 5.16(b). These intensities are recorded as I. The transmittance τ of the plume at each wavelength is given by

$$\tau = \frac{I}{I_0}$$

The IR channel is used to measure the effects of aerosols and particulate matter. The measurements at 400 and 600 nm superimpose the effects of NO_2 absorption, whereas SO_2 absorption is determined from the 310 nm measurement. This unit must be calibrated against known NO_2 and SO_2 concentrations. This can be done on an experimental stack in which sensors are inserted in the stack itself.

A basic radiometer for this mode of use consists of a lens system to focus the light on a broad-band optical detector. The output of the detector is amplified and connected to a metering circuit as indicated in Fig. 5.16(a). Rechargeable batteries can be used to power the instrument.

Dynamic response of this instrument is not very critical. One must avoid saturating the radiometer circuit when the instrument is pointed at the sun. Most optical detectors have a very rapid response time. The main error source in this system is the time delay between the background measurement and the plume measurement. This is the reason for the requirement for a cloud-free area around the sun, so that fluctuations in light intensity are avoided.

The device just described is really a very simple form of spectrophotometer (section 10.1) called a colorimeter, in which optical filters are used to select certain wavelength bands. Spectrophotometric techniques are used in a variety of instruments for determining chemical composition and optical density of materials. Biochemical processes in which optical densities change, such as enzyme reactions, are frequently monitored by these techniques.

The fundamental spectrophotometer system is shown schematically in Fig. 5.17(a). The basic system consists of a light source (the sun in the radiometer just discussed), an optical collimator and wavelength separator (usually lenses, slits, and diffraction grating, but sometimes a lens and filters), the sample, optical detector (phototube or photodiode), amplifier, and recorder/display system. Such instruments are necessarily large and rather cumbersome for field use, as well as being susceptible to damage. Alignment of the optical system is critical and it is sensitive to shock and vibration.

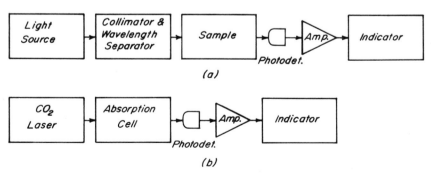

FIG. 5.17 Spectrophotometer systems: (a) basic optical system using conventional light source; (b) CO_2 laser spectrophotometer for gas analysis.

Many of the problems associated with spectrophotometers can be avoided by using lasers when single wavelength measurements are required. When correctly designed and adjusted, lasers generate a collimated (optically coherent) light beam of essentially single wavelength. Thus a laser replaces the light source, collimator, and grating in a conventional spectrophotometer system, at the expense of single wavelength versus adjustable wavelength operation. In many cases, this is not disadvantageous. Many small portable, and quite inexpensive laser systems are commercially available. Laser theory and design is well beyond the scope of this book; interested readers are referred to references such as Siegman, 1971, or Svelto, 1976.

To complete a portable instrument, one needs to add to the laser source only a sample chamber, a small optical detector, and an amplifier/indicator. Also commerically available are detector diodes matched to particular lasers.

Lasers are particularly well adapted to the detection of certain gases, such as CO_2. Carbon dioxide gas lasers produce an emission line at 10.6 μm, and a CO_2 detection system is illustrated in Fig. 5.17(b). Other gas lasers include ammonia, helium, neon, and various mixed gas systems. Since emission and absorption lines are identical, laser systems provide a convenient method for both qualitative and quantitative measurements.

Applications of laser systems will be examined in this and subsequent chapters.

5.12. PARTICLE CONCENTRATION METER

In both medical laboratories and water quality analysis, a device known as a nephelometer is used. This is basically a spectrophotometer (Chapter 10) in which one observes the light scattered by a sample, rather than the light absorbed. This instrument is used to determine lipids in blood serum and particulate matter in drinking water, among other applications. An improvement on this technique developed at NASA Goddard Space Flight Center (GSC-12088) correlates particle concentration in a fluid medium with particle composition; it is illustrated in Fig. 5.18.

Applications of this device include detection of relative amounts of organic and inorganic contaminants in river water, liquid industrial wastes, and other liquid media. With correct calibration, the contributions of various species to changes in contamination levels can be monitored.

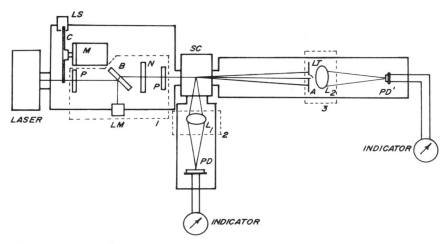

FIG. 5.18   System for determining particulates in liquids (redrawn from *NASA Tech. Briefs*, Fall, 1976): A = 2° scattering aperture; B = beam splitter; C = light chopper wheel; L_1 = lens; L_2 = lens; LM = laser output monitor; LS = light source and photodetector to monitor chopper operation; LT = light trap; M = light chopper drive motor; N = neutral-density filter; P = pinhole spatial filter; PD = 90° photoelectric detector; PD′ = 2° photoelectric detector; SC = scattering cell or diffusers used in calibration; dashed areas indicate separate optical systems.

The main feature of the instrument design is its ability to make a simultaneous measurement, in a liquid, of the light scattered at a small angle (2°) and the light scattered at a large angle (90°). A nephelometer is capable of measuring only large-angle scattering. By measuring both small- and large-angle scattering, the instrument distinguishes between high- and low-refractive index particles. Typically inorganic particles have a high refractive index and scatter light strongly at large angles. Organic particles generally have a low refractive index and scatter light through small angles.

As shown in the figure, the instrument has three optical paths. The first, consisting of pinhole spatial filters, beam splitter, neutral-density filter, and laser output meter forms an optical reference channel to monitor the output light intensity of the laser source. A chopper wheel placed between the laser source and the optical system is used to interrupt the emitted light periodically. The light source passed through the upper edge of the chopper is detected and used as a reference signal in the detection of the scattered laser light. The functions of the neutral density filter and the pinhole filters are, respectively, to prevent saturation of the photodetector and to stop stray scattered and reflected light from reaching the scattering cell.

The second optical path through a lens at 90° to the laser beam detects the large-angle scattered light. The third optical path, which consists of a focusing lens, circular aperture, radiant energy mask, and optical sensor, detects the light scattered from the rear of the scattering cell. The lens focuses the rays, scattered at 2° to the incident beam, onto the photoelectric detector. The two photodetectors are connected to separate recorders, connected for synchronous recording of the 90° and 2° scattering signals.

The dynamic response of this instrument is not critical since the measurement is static rather than dynamic. The sample, however, should be well mixed so that all of the particulates are in suspension.

5.13.   LASER PARTICULATE SPECTROMETER

Figure 5.19 illustrates a device proposed by the Lyndon B. Johnson Space Center for monitoring airborne particulate matter. The instrument consists of a laser light source, a sample chamber, and an opto-electronic detector system. The basic specifications for this system are:

Particle diameter	0.8–2.75 μm
Particle speed	0.2–20 m/s
Operating temperature range	77–300 K
Operating pressure	ambient to ultrahigh vacuum

A hybrid laser scattering-and-extinction technique is the basis for instrument operation.

The particles are directed through the laser beam within the laser's resonant cavity. The laser is operated in a mode such that the

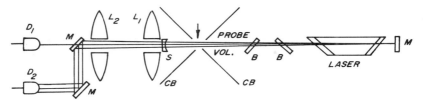

FIG. 5.19   Laser particulate spectrometer (redrawn from *NASA Tech. Briefs*, Fall, 1976): B = compensating Brewster windows; CB = conical baffles; D_1 = scattering photodetector; D_2 = extinction photodetector; L_1 = first scattering collection lens; L_2 = second scattering collection lens; M = plane mirror; S = spherical laser resonator mirror; arrow points to probe volume.

cross-section of the beam is a dark center area surrounded by a bright ring of light. As the particles pass through the beam, they scatter light away from the beam and hence reduce its effective intensity. Two photodiode detectors are used: one detects the scattered light while the other detects the reduction in intensity of the main beam. The particle-size-range limitation results from the fact that particles greater than 10 μm in diameter severely perturb or extinguish the laser beam. Very small particles are not detected.

The instrument employs a single He–Ne laser tube enclosed in a hermetically sealed chamber for mechanical protection and thermal regulation. The laser beam exits through a double Brewster-angle window so that beam alignment is maintained. The laser is maintained at constant temperature.

The signals produced by the two photodetectors are proportional to the number of particles present. The data are analyzed using an electronic multichannel analyzer and calculator. Calibration of the instrument is achieved by spraying particles of known size and glass microbeads through the beam. The dynamic response of the instrument is limited only by the response time of the photodetectors. Particle size limitation has been discussed above.

5.14. PORTABLE DRUG DETECTION SYSTEM

A current problem is rapid screening and detection of suspected drug compounds. The Ames Research Center of NASA has developed a portable field instrument for this purpose. The instrument carries US Patent No. 3,814,939. Figure 5.20 illustrates the basic system, which is a special purpose spectrophotometer (see Chapter 10). A suspected drug sample is chemically treated so that it becomes highly fluorescent. It is then placed in a chromatographic column of similar composition (solvent and packing material) to the thin-layer chromatography techniques normally used for drug identification. In the column, the samples separate into narrow bands.

The column is irradiated with ultraviolet light, which causes the narrow sample bands to fluoresce strongly at wavelengths characteristic of the particular sample. The visible light thus reflected from the column is introduced into a collimator and reflection grating. Light from the grating is focused on a photodetector. The electrical signal from the detector is amplified and displayed on a strip chart recorder. Each

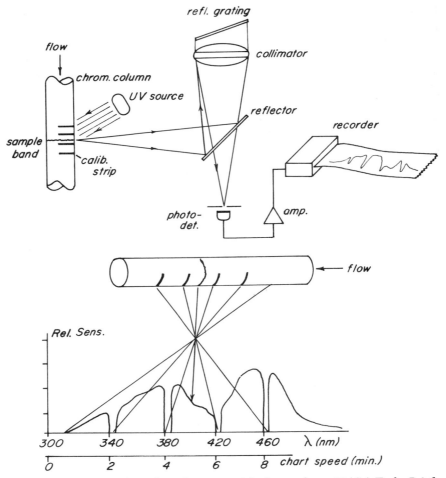

FIG. 5.20 Portable drug detection system (redrawn from *NASA Tech. Briefs*, December, 1976, ARC 10633).

compound produces a characteristic spectral signature on the chart record. The dark areas between the fluorescent bands provide both wavelength calibration and timing data as the compounds move through the chromatograph column. The unit is both self-scanning and self-calibrating. The only moving part is the chart recorder.

A major application of this instrument is in determining drugs such as morphine in urine samples. The urine sample is chemically treated to convert morphine, which fluoresces weakly, into a strongly fluorescent fluorophore. After this chemical treatment, the specimen is introduced into the chromatography column.

Since urine is an aqueous solution, the fluorophores produced should be those which fluoresce strongly in aqueous media. Some appropriate fluorophores are: pseudomorphine, dansylmorphine, and the reaction products of morphine with such compounds as Marquis reagent, 3-phenyl-7-isocyanatocoumarin, and the diazonium salt of aminorosamine.

To implement the self-scanning and self-calibrating aspect of this system, the chromatography column is equipped with a number of masking strips along its length. These strips are equally spaced along

Table 5.1

Noise Sources and their Typical Sound Pressures

Sound pressure, bar $\times 10^{-6}$	Sound level, dB		Source
		140	
1000	134		Threshold of pain
		130	
			Pneumatic chipper
		120	
100	114		Loud automobile horn (dist. 1 m)
		110	
		100	
10	94		Inside subway train (New York)
		90	
			Inside bus (internal combustion engine)
		80	
1	74		Average traffic on street corner
		70	
			Conversational speech
		60	
0.1	54		Typical business office
		50	
			Living room, suburban area
		40	
0.01	34		Library
		30	
			Bedroom at night
		20	
0.001	14		Broadcasting studio
		10	
			Threshold of hearing
0.0002		0	

the column length and should be as wide as the maximum anticipated width of any fraction band as the compounds move through the column.

The masks and the associated column are always in a fixed position relative to the optical system. The position of each masking band corresponds to a particular wavelength relative to the optical system and grating. Interruptions of the time spectral trace when any given fluorescent band, while moving through the column, is obscured by each mask provides a spectral calibration point that relates chart position and spectral wavelength directly. Since the distance between masks and the recorder chart speed is accurately known, the rate of progression of the fluorescent band through the column can be calculated. Quantitative measurements are obtained from the following data: fluorescent spectrum characteristic of sample, spectral signal amplitude, and rate of movement of the component band through the column.

5.15.   NOISE MEASUREMENTS

The human ear is a very sensitive organ and there are now legal requirements relative to the Federal Occupational Safety and Health Act that sound levels be monitored in many industrial operations. The ear itself is a frequency-sensitive transducer that responds nonuniformly over the range 20–15,000 Hz. The physical unit by which sound is

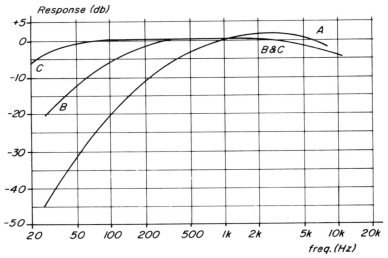

FIG. 5.21   Weighted frequency-response scales for sound level meters.

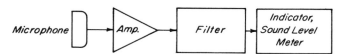

FIG. 5.22   System schematic for sound level meter.

measured is pressure, usually expressed in a logarithmic unit, the decibel (dB). Since the dB is dimensionless, sound level is actually measured relative to a reference sound pressure. This reference is $0.0002 \, \mu$bar $(2 \times 10^{-5} \text{N/m}^2)$. Table 5.1 illustrates some typical sound sources and their equivalent dB values.

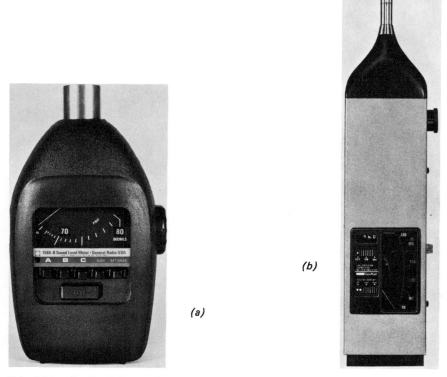

(b)

(a)

FIG. 5.23   GenRad (formerly General Radio Co.) sound level meters: (a) model 1565-B portable meter. (b) Model 1982 precision sound level meter and analyzer. Photographs courtesy of GenRad, Inc., Concord, MA.

A device to monitor sound level (sound pressure) must be designed so that it "hears" in the same way as the human ear with respect to frequency response. Figure 5.21 shows the different response scales that are used in the design of sound level meters. The "A" scale is normally used. The "ear" of the device is a piezoelectric crystal. The output of the crystal is amplified and electronically filtered to give an output that replicates the "A," "B," or "C" frequency characteristic. The system is completed by a meter that is calibrated in dB. A system schematic is shown in Fig. 5.22 and typical devices in Fig. 5.23.

Table 5.2

Term	Unit	Definition	Use
Candlepower	Candela, cd	International basic physical quantity for all light measurements	Defines luminous intensity of a source in one specific direction
	Candlepower, cp	Candlepower is the luminous intensity of a source, expressed in candelas	A property of a light source which defines luminous flux at the source origin
Luminous flux	Lumen, lm	The light flux falling on a 1 ft² surface, of which every point is 1 foot from a point source of 1 cd in all directions	Total light source output, amount of reflected light, light aborbed or transmitted by an object, amount of incident light falling upon a surface
Illumination	Footcandle, fc	Illumination at a point on a surface 1 ft from and perpendicular to a uniform source of 1 cd. Luminous flux density on a surface	Indicates illumination at a specific point, or average illumination on a surface
Luminance (photometric brightness)	cd/in², footlambert, fl	Brightness of an illuminated surface— cd/unit area or fl/unit area. Emitted or reflected light from a surface at the rate of 1 cd/in² of projected area = luminance in that direction of lcd/in²	Convenient unit for expressing luminance of illuminated surfaces

5.16. LIGHT MEASUREMENT

The basic principle of most light meters lies in the photovoltaic effect; that is, a voltage is generated when electromagnetic energy (light) strikes a receptor unit. Usually cadmium sulfide (CdS) cells are used. The electrical output from the cell is connected to a microammeter that is calibrated in one of the light intensity units, usually footcandles. Table 5.2 indicates the various units that are in common usage. Figure 5.24 shows a schematic for two types of instruments and a field unit is shown in Fig. 5.25. For industrial use, meters must be calibrated both for light color and the angle at which light enters the receptor. This latter correction is called cosine correction.

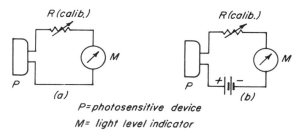

P= photosensitive device

M= light level indicator

FIG. 5.24   Circuits for photosensitive meters.

FIG. 5.25   General Electric Co. portable photometer with color and cosine correction (courtesy J. T. Banner Associates, Inc., Laramie, Wyoming).

5.17.   SUMMARY

This chapter has presented some practical transducers used in clinical and environmental applications. In the next chapter, we examine electrodes and electrode systems that find extensive application in many scientific fields.

5.18.   REFERENCES

Cobbold, R. S. C., 1974, *Transducers for Biomedical Measurements, Principles and Applications*, Wiley, New York.

Geddes, L. A., and L. E. Baker, 1975, *Principles of Applied Biomedical Instrumentation*, Wiley, New York.

Siegman, A. E., 1971, *An Introduction to Lasers and Masers*, McGraw-Hill, New York.

Svelto, O., 1976, *Principles of Lasers*, Plenum, New York.

Thomas, H. E., 1974, *Handbook of Biomedical Instrumentation and Measurement*, Reston Publishing Co., Reston, Virginia.

NASA *Tech. Briefs*, 1975-77. (Includes reference GSC-12088).

6

Electrodes

6.1.   INTRODUCTION

Electrodes are one of the most common form of transducer used in clinical, physiological, and biological applications. They are used as interfacing media between electronic instrumentation and living systems. In this mode, they may be used to stimulate the living system, or they may be used to detect electrical activity produced by such a system. Familiar clinical applications are electrocardiogram (ECG) and electroencephalogram (EEG) recording. For applications of this nature, metal electrodes are normally, but not exclusively, employed.

Electrodes find many applications in chemical analysis, industrial process control, and environmental monitoring. In these areas, ion-specific electrodes are used extensively to detect the presence of chemical entities. Normally these electrodes are one-half of an electrochemical cell; the other half is a silver–silver chloride or calomel reference cell. Rather than being metal, ion-specific electrodes have as their active sensor a glass membrane, or a membrane of some other material. The glass pH electrode is a familiar example of an ion-specific electrode.

In this chapter, we explore some of the electrode techniques that may be applied in instrumentation systems. Metallic electrodes, because of their simplicity will be examined first, and then we will discuss ion-specific and other types.

6.2. METALLIC ELECTRODES

Metal electrodes are used extensively for both recording and stimulating in biological systems. They come in many forms and are fabricated from many types of materials. There are several general design criteria that must be respected and appreciated. Generally speaking, the metals or alloys used should be as chemically inert as possible (corrosion resistant) and nontoxic. Typical materials are the noble metals, stainless steel, German silver, and tungsten compounds. In addition to low chemical activity, the mechanical strength of the metal may also be important, particularly in the case of microelectrodes. For most applications, copper should be avoided because it is toxic and corrodes easily in the presence of chloride ions associated with physiological fluids.

Normally, metal electrodes are used to record natural or evoked potentials associated with living systems. Primary applications are the diagnostic electrocardiogram shown in Fig. 6.1 and in the electroencephalogram shown in Fig. 6.2. Both recording techniques have standardized electrode locations. One such system, illustrated in Fig. 6.3, is known as the Einthoven triangle system for the pioneer of ECG recording. Figure 6.4 shows the designations for the events which occur during the cardiac cycle. The P wave represents atrial contraction. The QRS complex is the electrical activity produced by ventricular contraction and repolarization of the atrium. The T wave represents repolarization of the ventricles. The origin of the U wave is not clear. Repolarization of the fibers in the bundle of His or the aorta have both been cited as possible sources.

Electrodes used for surface recording, such as ECG records, are typically fabricated from the alloy German silver or stainless steel. For EEG recording, gold electrodes may be used or silver–silver chloride pellet electrodes. In some cases, subcutaneous stainless-steel needles are used. Electromyographic (EMG) recording and stimulation procedures generally make use of needle electrodes. Either stainless steel or platinum–iridium may be used, and sometimes tungsten. Surface electrodes are usually discs or oblong in shape and are frequently contoured to fit the portion of the body to which they are

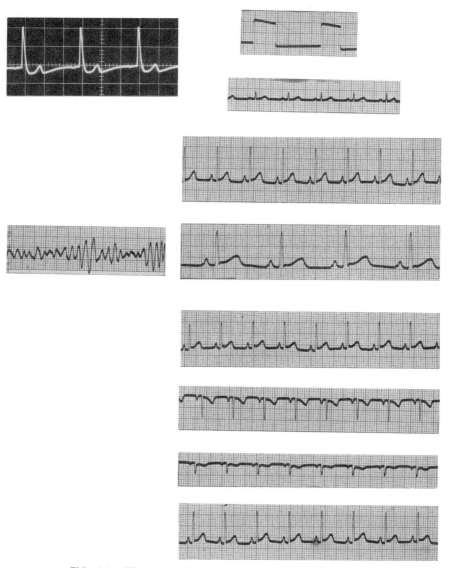

FIG. 6.1   Electrocardiogram signals from a canine heart.

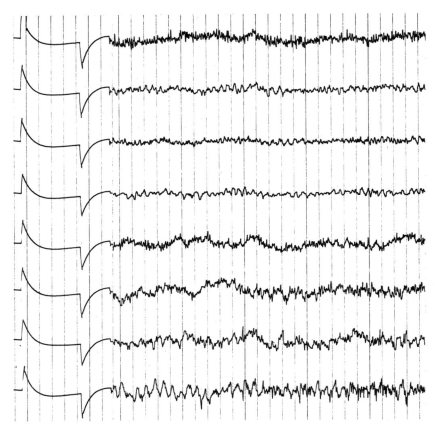

FIG. 6.2   Typical electroencephalographic records from a normal human subject. The traces on the left are 50 mV calibration signals that also indicate the dynamic response of the amplifiers. (Courtesy of Dr. John W. Steadman, University of Wyoming).

applied. They are secured by means of adhesive tape, elastic bands, or in some cases by a suction bulb. When the latter technique is employed, one must take care not to produce a suction strong enough that surface blood vessels are ruptured. Care must also be taken such that the electrodes are not too tightly attached when elastic straps are used, as excessive pressure can evoke EMG signals that appear as "noise" in the potentials being recorded.

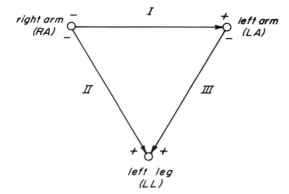

FIG. 6.3 Polarity convention for Einthoven triangle: leads I, II, and III. The terminal polarities shown produce positive recorder deflection. The arrows indicate the sense of current in the patient to produce positive recorder deflection.

6.3. ELECTRODE INTERFACE CONSIDERATIONS

The interface between an electrode, but especially a metallic one, and an electrolyte, is extremely critical. Generally it is an electrode–electrolyte interface, where either a biological tissue or a wetting solution is the electrolyte. Silver–silver chloride electrodes require the use of a wetting solution or paste which contains Cl⁻ ions; otherwise the electrodes will not operate correctly. A metal electrode–electrolyte interface

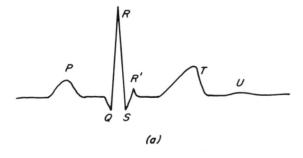

(a)

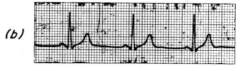

(b)

FIG. 6.4 Designation of ECG events: (a) idealized ECG waveform showing the components; (b) lead II tracing from young adult human male.

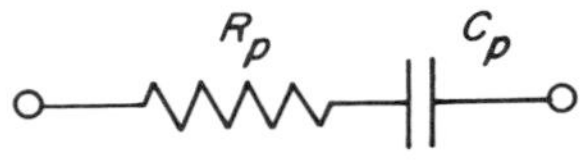

FIG. 6.5   Representation of AC electrode polarization impedance.

poses several problems. The electrochemical activity that occurs at such an interface results in an electrical potential difference across the interface. This potential difference produces a DC offset that can influence amplifier systems. If alternating current is passed through the interface, an additional frequency-dependent polarization phenomenon is observed. It can be represented as a series R–C circuit, as shown in Fig. 6.5, where the R and C are both frequency and current density dependent. Below a certain threshold current density, R and C are linear and vary only with frequency. Above this threshold, R and C vary with current density. If we define an AC electrode polarization impedance

$$Z_p = R - j/\omega C$$

and the DC polarization is defined as V_0, then current density threshold is approximately

$$J\text{th} > V_0/S|Z_p|$$

where Jth is the threshold current density (A/m^2), S is the surface area of interface (m^2); and $|Z_p|$ is the magnitude of the AC polarization impedance (ohms). This condition is illustrated graphically in Fig. 6.6.

The effects of an interface impedance or polarization are multifold. A potential drop occurs which can be significant in some applications; electrical noise may be generated; pulsatile or nonsinusoidal signals may be severely distorted. There is no way to eliminate the AC polarization impedance when electrodes are used for stimulation, since, by necessity, a current density must exist at the electrode–electrolyte interface. When electrodes are used in a recording configuration, the interface impedance can be reduced substantially by using very high input impedance amplifiers so that very small signal currents exist.

Since AC polarization depends upon current density, one way to decrease the effect would be to increase electrode surface area. In most cases, this is not practical. The effective surface can be enlarged, however, by roughening. A convenient way to do this is by platinization, which is most effective for platinum or stainless steel. Platinizing is a technique whereby colloidal platinum black is deposited on the elec-

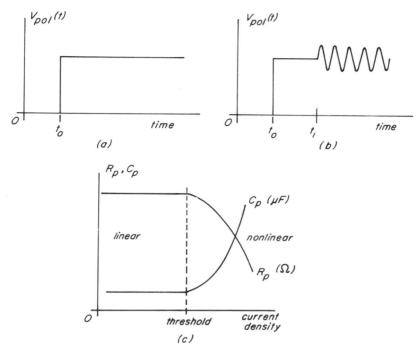

FIG. 6.6   Electrode polarization phenomena: (a) DC polarization; (b) DC elec-trode polarization with superimposed AC polarization potential; electrode–electrolyte contact is made at time $t = t_0$, and the AC potential is applied at time $t = t_1$; (c) behavior of C_p and R_p as a function of current density at the electrode–electrolyte interface.

trode surface. A similar technique is used to plate colloidal gold on gold electrodes. The resulting surface has a velvety aspect and, with care in technique, produces an effective surface area that may be several orders of magnitude larger than the geometric area, as a result of the granularity of the coating.

The platinizing technique is somewhat of an art and rather com-plex if one is to assure repeatable results. The electrodes to be treated must be absolutely free from grease and other contaminants, and should be sandblasted before plating. A bright platinum electrode is used as the anode and the electrode to be plated as the cathode. Best results are obtained with plating current densities on the order of 10 mA/cm². Improved coatings are obtained by interchanging the anode and cathode connections for short periods of time. The regimen is somewhat tricky and is discussed below. Platinization effectively increases C_p on the series basis, thus it decreases the term $(j\omega C_p)^{-1}$, and it decreases R_p so that Z_p is decreased.

When pairs of electrodes, such as would be used in an impedance bridge measuring cell, are platinized, the electrodes, after preparation, should be shorted together in a saline solution and left for several days so that the residual potential difference produced by plating can be equalized.

Platinization is accomplished by the use of conventional electroplating techniques. The electrode to be platinized is connected as the cathode, and a piece of pure platinum sheet forms the anode of an electrolytic cell. Kohlrausch and Holborn formula platinizing solution is employed as the electrolyte. Its chemical composition is 3% platinum chloride (H_2PtCl_2) dissolved in a 0.025% lead acetate solution. This platinizing solution may be obtained in 2-oz. bottles from the Hartman-Leddon Company in Philadelphia or from the Arthur H. Thomas Company, also of Philadelphia.

Prior to platinization, the electrodes must be cleaned and the surface prepared. The electrode surface should first be sanded with a fine emery paper and water to remove previous colloidal platinum coatings. Boiling in *aqua regia* may be necessary. The electrode surface is then sandblasted with American Optical Company emery #M180, followed by washing with distilled water. The water wash is followed by three rinses: 1. in acetone, $(CH_3)_2CO$; 2. in a solution of two parts by volume of ethyl alcohol 95%, C_2H_5OH, 1 part acetone, 1 part methyl alcohol, CH_3OH; 3. absolute ethyl alcohol. Following the conclusion of this preplatinizing surface treatment, the electrodes are now placed in a platinizing cell which holds them in place during plating. It is rinsed with spent platinizing solution and filled with new platinizing solution. Plating cells can be constructed with two sections. This permits simultaneous platinizing of a matched pair of flat electrodes. A current of 10 mA/cm² is passed through each section of the cell for 25 min. The cell is then emptied, the electrodes rotated 180°, the cell refilled, and the plating carried out for an additional 25 min. Rotation of the electrodes prevents the formation of a thickness gradient in the coating. A good coating should have a fine-grained velvety black appearance.

Figure 6.7 illustrates the effect of platinizing time upon the series electrode polarization capacitance with current density as a parameter. Very low current densities, although providing a good surface, require a prohibitively long plating duration. On the other hand, large current densities cause flaking off of the deposited platinum particles. A

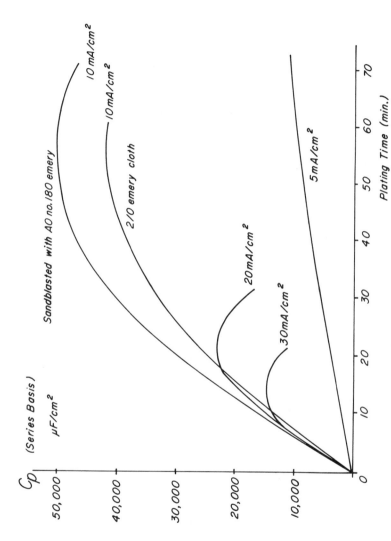

FIG. 6.7   Dependence of electrode polarization capacitance on plating time, current density, and surface treatment. Electrode area = 5 cm².

current density of 10 mA/cm² appears to be the optimum value. A 0.9% saline (NaCl) solution is taken as the electrolyte for the C_p determination with an electrode separation of 1.0 cm.

After the electrodes have been prepared, they should be placed in a suitable holder that is filled with distilled water or physiological saline solution. The electrodes are then connected together to form a short circuit, and should remain this way for several days until the charge produced by the platinizing procedure has decayed.

A few general comments concerning the care of platinized electrodes are in order. The electrodes should be stored in distilled water. Before use, they should be allowed to stand for 30 min in contact with the material to be investigated. If the electrodes have not been damaged during use, replatinizing is unnecessary. The electrodes should be removed from the measuring cell, washed with distilled water and ethyl alcohol, and replaced in distilled water. If grease accumulates on the electrodes, it should be removed immediately by washing with ether. Electrodes should be replatinized if grease cannot be removed or if the electrode surfaces are damaged or generally worn down. If a heavy accumulation of grease forms, the electrodes should be placed in boiling *aqua regia* for 5 min.

When electrodes are used for impedance measurements, zero-current techniques should be used rather than conventional AC bridge measurements. Inherent in bridge operation is the passage of current through the active bridge impedances. Although electrodes can be platinized, this method is not a panacea. It extends the lower frequency range over which electrolyte impedance measurements can be made, and reduces the effects of AC electrode polarization impedance. The only way in which the polarization impedance can be eliminated is to make measurements in a manner that eliminates the current density at the electrode–electrolyte interface.

The simplest zero-current method is the basic four-electrode system shown in Fig. 6.8. The specimen to be measured forms one arm of an electronic half-bridge and supports a series current. The measuring circuit consists of two virtually zero-current potential probes. Bridge balance is achieved when the decades R_v and C_v are adjusted so that the voltage across this combination is equal in magnitude and phase to the voltage developed across the potential electrodes of the measuring cell. The voltage divider is required to produce an effective voltage gain of unity in the first amplifier.

With the system shown, a resistance resolution of 1 part in 100,000 is possible at 1 kHz, which degenerates to 1 part in 1000 at 1 Hz. System noise and detector inadequacies account for loss in resolution

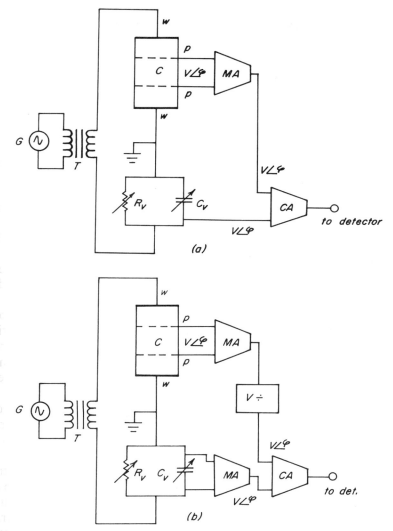

FIG. 6.8 Zero-current 4-electrode techniques: (a) simplified system where *MA* is set to unity gain; (b) balanced amplifier 4-electrode half-bridge system.

C is the sample chamber, *CA* is the comparator amplifier, *MA* is the measuring amplifier, C_V is the balance capacitor, R_V is the balance resistor, *T* is the bridge transformer, *G* is the generator, *w* are the working electrodes, and *p* are the probe electrodes.

These systems are susceptible to stray capacitances unless care is taken during use. The electrical connections between *MA* and the sample chamber must be kept as short as possible.

Capacitance resolution is governed by the relation

$$\frac{\Delta R/R}{\Delta C/C} = \omega RC$$

where R and C are the true values of the sample.

This system has one shortcoming. Unless the two amplifiers are very carefully matched so that no relative phase shift exists, capacitance measurements are meaningless although resistance determinations can be made. If the relative phase shift is small, one may substitute a parallel R–C circuit for the potential electrodes of the cell and rebalance for null after initial balance of R_v and C_v. The settings of the new R–C decades are the true values for the electrolyte. The system has the advantage that direct readings are obtained with minimum time and effort. The design of the sample cell is somewhat critical and requires some thought relative to the nature of the desired measurements. The amplifiers used must have high common mode signal rejection.

It should be noted that the so-called nonpolarizable electrodes, such as silver–silver chloride, are misnamed. There is no such entity as a nonpolarizable electrode from an electrochemical point of view.

6.4.   MICROELECTRODES

Microelectrodes fall into two general categories, metal and glass pipet. They are distinguished from gross electrodes by the fact that active electrode surfaces are small enough to contact a single cell or neural unit. Electrode tip sizes generally lie in the 0.25–5 μm range. In shape, microelectrodes are usually tapered needles in order to reconcile a small tip diameter to a reasonably large diameter shaft. The latter is necessary for mechanical positioning of the electrode and for making necessary electrical connections to the recording system or excitation source. Figure 6.9 illustrates the nomenclature associated with microelectrodes.

In terms of the mechanics of assembly, metal microelectrodes are the simplest to produce. They are simply metal wires or needles that

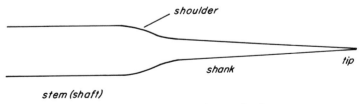

FIG. 6.9   Microelectrode terminology.

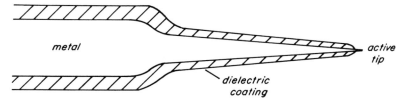

FIG. 6.10   Metal microelectrode (longitudinal section).

have a very small tip. All of the electrode, save the tip, is insulated with a suitable material, as shown in Fig. 6.10. The only mechanical problem associated with such electrodes is finding a stiff enough metal to insure electrode rigidity and suitable materials and techniques for providing the insulating sheath.

Electrodes can be made by firmly attaching a stiff wire to a larger diameter shaft, or by drawing rod stock through a die. The shaft end is fitted to an appropriate electrical connector. The tip end requires careful finishing before the electrode can be used. Normally, the tip is first ground smooth. Grinding rather than cutting is preferred since this prevents the tip from fraying. The final pointing of an electrode tip is done by electrolytic methods. The exact technique depends upon the electrode material, but usually involves etching or electropolishing.

After the metal microelectrode has been fabricated, it must be insulated except for the tip. Various materials can be used. Dielectric coating is normally accomplished by dipping the needle into the coating material and slowly removing it. For any particular insulation, one must experiment at first in order to develop a good technique for the particular electrode and insulating materials being used. Usually several coats of thinned insulating material work best. The diluted coating material provides uniform insulation and will draw back from the tip to leave it exposed. Some coatings, such as Formvar, require baking (30 min at 120°C) after coating. Drying in a gentle flow of warm air is usually adequate for lacquers.

Many of the electrodes used in electrophysiology work are fluid-filled glass pipets. The basic structure is illustrated in Fig. 6.11. The electrode is a glass capillary drawn to a fine point. The lumen is filled with an electrolyte, usually an aqueous KCl solution, and a metal wire is inserted in the stem to form electrical connection.

Of the three commonly available types of glasses, the borosilicate (Pyrex-type) glasses appear to be the most satisfactory. They combine the properties of good electrical resistance, resistance to thermal shock, and mechanical strength. The soda–lime glasses have poor heat resistance

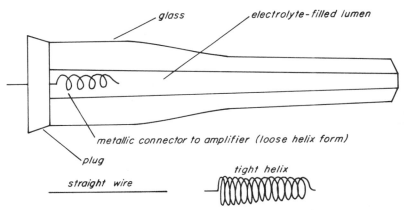

FIG. 6.11   Glass micropipet and methods for metallic connections.

and chemical stability, whereas the lead glasses tend to be fragile in electrode work and also possess a high temperature coefficient of expansion that makes them subject to thermal shock.

6.4.1.   Electrode Pulling

Many types of electrode pullers are commercially available and individual scientists often construct their own units for particular applications. Figure 6.12 illustrates the basic device. Both vertical and horizontal pullers are available, and a vertical assembly is shown in Fig. 6.12. The device consists of a rigid vertical support and base. At the top, a fixed clamp is located to hold the glass capillary tubing firmly. The tubing is positioned through a heater element and a mass is clamped to the bottom end of the tube. The heater temperature and the mass must be adjusted experimentally for the type of glass used. With proper adjustment of these parameters, the capillary tube will neck down and separate cleanly leaving an open tip 0.5–1 μm in diameter. A simple mass acting under the influence of gravity can be used. More sophisticated devices use a spring and solenoid arrangement in place of a mass. In this case, a very lightweight clamp is used on the bottom end of the tube. Adjustable tension springs and variable pull solenoids are used to achieve optimum pulling conditions. For a given formula glass, the two variable system parameters are heater (filament) temperature and spring tension (effective mass). The heaters used are generally platinum foil or filaments. Temperature control is obtained by operating the filament from a low-voltage high-current transformer (such as one used for spot welding) which in turn is controlled by an autotransformer (variac). The

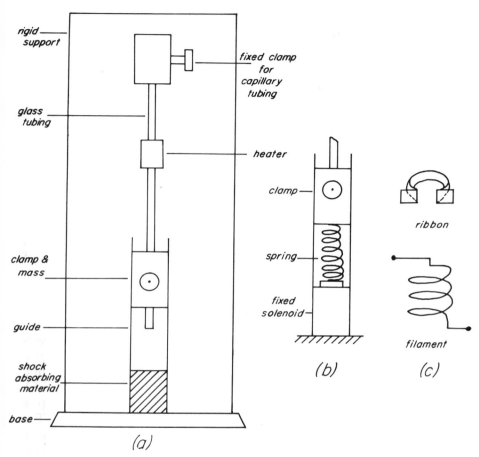

FIG. 6.12 Simple microelectrode puller: (a) main unit; (b) alternative "mass" spring–solenoid system; (c) two types of heater units.

reproducibility of micropipets depends upon constancy of pulling tension, heater temperature, and heater position. The filament or foil must be secured rigidly to prevent sagging. If the filament sags during pulling, uneven electrodes may occur, or too long a length of glass may be heated, producing an undesirably long shoulder on the electrode. Open wire filaments are the most susceptible to sagging. It is better to support wire filaments rigidly in a tube of refractory material such as transite.

With proper cushioning of the mass or solenoid assembly after fracture of the glass has occurred, it is possible to produce two electrodes simultaneously, one in the fixed clamp and the other attached to the mass or solenoid unit.

The length of the heater and, to some extent, the pulling force, control the geometry of the finished micropipets in regard to shoulder taper and shank length.

6.4.2.   Pipet Filling

A problem exists in filling micropipets, especially those with tip diameters less than 5 μm. Larger sizes can be filled with electrolyte introduced from a syringe via a hypodermic needle. The needle is pressed against the glass inside wall of the pipet's tapered section. An alternative technique is to use another micropipet in place of the hypodermic needle. A unit with approximately 5 μm tip diameter is nested inside the pipet to be filled. Filling operations of this sort should be carried out under a microscope to insure complete filling of the micropipet lumen and to prevent damage to the pipet tip.

Individual handling in this manner of very small pipets is not recommended since tip damage occurs too easily. Various methods have been proposed for automatic filling. Basically these involve putting one or several pipets into a suitable holder, which is then immersed in the filling electrolyte, and boiled for several hours. Boiling time and subsequent tip-erosion damage can be reduced if the filling solution is first heated and the assembly placed in a vacuum.

An alternative method involves placing the pipets into 40°C methanol and then inserting the assembly into a vacuum chamber. The chamber is evacuated and the alcohol allowed to boil for approximately 8 min, during which the micropipets are filled. The assembly is then removed from the vacuum chamber and the methanol replaced by the filling electrolyte. In about two days, the methyl alcohol will be replaced with electrolyte by diffusion. The disadvantage of this process is the two day delay.

Ideally the filling electrolyte should be a saline solution which is isotonic to the physiological system in which the microelectrode is to be used. In practice, with such solutions, the resulting electrode resistance is too high. For this reason a $3M$ KCl electrolyte is frequently used; other possibilities include 1% NaCl, $2M$ NaCl, $0.6M$ K_2SO_4, and so on. In some instances, the electrode application dictates which electrolyte should be used.

To prevent electrode plugging, freshly prepared and filtered electrolytes should be used. They should be sterilized by boiling to reduce bacterial growth in the micropipet lumen and contamination of the life system being studied.

Generally, micropipet electrodes should be stored in distilled water or alcohol. If stored in the filling electrolyte, bacterial growth may occur, and in any event strong electrolytes tend to erode fine tips. Storage in the dark at low temperature retards bacterial growth. Micropipets stored in and filled with distilled water or alcohol can be filled with electrolyte for use by the diffusion technique mentioned above. Electrolyte-filled electrodes stored in air are frequently damaged by crystallization of salts on their tips, with resultant tip fracture.

Electrical connection is made to glass microelectrodes by the methods shown in Fig. 6.11. Either bare wire or chlorided silver wire is used. Microelectrodes also manifest electrode polarization effects because there is a metal–electrolyte interface. Serious signal distortion can occur if current is passed through a microelectrode. Normally the effective electrical resistance of a glass micropipet is of the order of 50–100 MΩ. This means that very high input impedance amplifiers must be used. Because any amount of stray capacitance can seriously distort electrode signals, negative-input capacitance amplifiers are frequently used to compensate for this effect. These amplifiers, however, do not compensate for interface capacitance and other capacitances associated with the electrode itself.

Because of electrode polarization effects, microelectrodes cannot be treated as simple conductors. The AC equivalent circuit for a metal microelectrode represents a high-pass filter. The network representation for a glass micropipet is a low-pass filter. These properties restrict, to some extent, the applications of these devices. Glass pipets are best suited for recording slow events, such as resting and action potentials from single cells. Metal microelectrodes are suited to recording fast events as associated with neurological signals. A generalized microelectrode recording system and equivalent electrical network representations are shown in Fig. 6.13.

6.5.   ION-SELECTIVE AND REFERENCE ELECTRODES

When a metal is in contact with an electrolyte solution, a DC potential occurs which is the result of two processes. These are 1. metallic ions which go into solution from the metal, and 2. the recombination of metal ions in the solution with free electrons in the metal to form metal

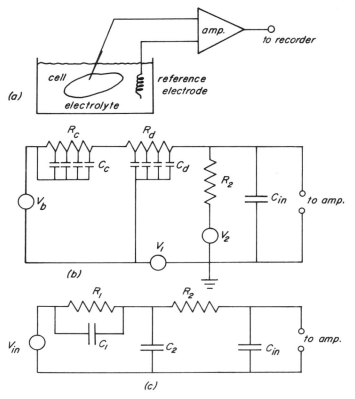

FIG. 6.13 Microelectrode recording system: (a) system arrangement; (b) and (c) electrical network representations for (a). $R_1 = R_c + R_d$ is the resistance of the electrode tip region; $C_1 = C_c$ is the capacitance between the cellular fluid or external electrolyte and the electrode tip region; $C_2 = C_d$ is the capacitance between the electrode shaft and an external electrolyte; R_2 is the shaft lumen resistance; V_b is the biological source voltage; and V_1 and V_2 are interface potentials.

atoms. After a metal electrode is introduced into an electrolyte, equilibrium is eventually established and a constant electrode potential is established (for constant environmental conditions). At equilibrium, a dipole layer of charge (electrical double layer) exists at the metal–electrolyte interface. There is a surface layer of charge near the metal electrode and a layer of charge of opposite sign associated with the surrounding solution. Although diffuse, this dipole layer produces an effective electrical capacitance (C_p) that accounts for the low-frequency behavior of the electrode polarization impedance as discussed above.

When a metal is in contact with an electrolyte that contains its ions, the electrode potential developed is called a half-cell potential since the configuration with one electrode represents half of an electrolytic cell. The table below indicates half-cell potentials for those metals that can be used in physiological systems (at 25°C).

Metals	Voltage
$Al = Al^{3+} + 3e^-$	+1.67
$Fe = Fe^{2+} + 2e^-$	+0.441
$H_2 = 2H^+ + 2e^-$	0.000 (reference)
$Ag = Ag^+ + e^-$	−0.7996
$Pt = Pt^{2+} + 2e^-$	−1.2
$Au = Au^+ + e^-$	−1.68
$Au = Au^{3+} + 3e^-$	−1.42

In addition to metal–electrolyte interfaces, electrode potentials can be produced by ion transport through an ion-selective semipermeable membrane. In this case, the membrane is interposed between two liquid phases. Reversible transfer of a selected ion occurs through the membrane. For an ideal membrane, the developed electrode potential, E, is given by the Nernst equation:

$$E = -\frac{RT}{ZF} \ln \frac{C_1}{C_2} \quad \text{Volts}$$

where F is the Faraday constant (96,495 C/mol), R is the gas constant (8.315 kJ/kg mol deg K), T is the temperature (K), Z is the valence of the ion involved, and C_1 and C_2 are the concentrations of ions on either side of membrane.

The Nernst equation predicts E accurately only in dilute solutions. For solutions normally encountered in physiological work, ion activities are used in place of ion concentrations. Activity, α, is defined by

$$\alpha = C\gamma$$

where C is ion concentration and γ is the activity coefficient for a given ion. In dilute solutions γ approaches unity and $\alpha \sim C$. The Nernst equation as modified by ion activities is thus:

$$E = -\frac{RT}{ZF} \ln \frac{\alpha_1}{\alpha_2}$$

In practice, if we fabricate an ion-specific electrode by using a semipermeable membrane, the voltage (emf) E that we measure is dependent upon α_1 and α_2. If we wish to use the electrode to determine ion concentration C_1 or C_2, then we must know the appropriate activity coefficients γ. These are evaluated through the Debye–Hückel equations, or we may simply calibrate the electrode using known standard solutions.

Half-cells and reference electrodes are generally categorized as reversible or nonpolarizable. By this, we mean simply that the electrode passes electric current without changing the chemical environment in the region of the electrode. The basic thermodynamic consideration for electrochemical reversibility is expressed by the Gibbs free-energy loss relation

$$\Delta G = -nFE$$

for an electrolytic cell consisting of a metal–metal ion half-cell electrode and an ideal reference half-cell electrode. Here, E is the cell emf, nF is the "capacity factor," and n is the number of faradays, F, transferred when the reaction proceeds to the amount $|\Delta G|$.

The basic relation which conerns us is:

$$\text{oxidant } (Z) + (ne) \rightleftarrows \text{reductant } (Z - n)$$

where Z is the valence, n is the number of electrons transferred per mole, and e is the electron charge. This relation expresses valence changes and the accompanying electron transfers that occur in oxidation–reduction systems. The basic property of such systems is the conversion of chemical energy into electrical energy.

6.5.1. The Silver–Silver Chloride Electrode

The silver–silver chloride electrode is based upon a reversible electrochemical reaction at the electrode surface. If the electrode is operated as one-half of a cell when it is the anode, Cl is deposited on the electrode; when it is the cathode, AgCl on the electrode surface is reduced to Ag and Cl^- ions are freed into the electrolyte solution. It consists of a metallic silver substrate, coated with AgCl, and is in contact with an electrolyte solution that contains a soluble chloride such as NaCl or KCl. Ag–AgCl electrodes are reversible or "nonpolarizing," although they do exhibit AC polarization effects. This means that the electrode can pass electric current without changing the chemical environment in the vicinity of the electrode. Since AgCl is slightly soluble in H_2O, the electric currents at the electrode–electrolyte interface must be kept relatively small to maintain an unchanged chemical environment.

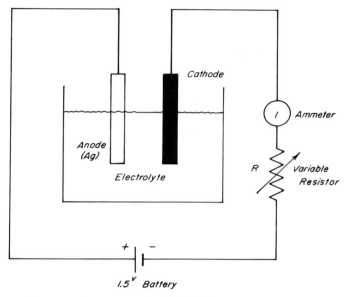

FIG. 6.14   Apparatus for chloridizing a silver electrode.

6.5.2.   Fabrication of Ag–AgCl Electrodes

Various techniques exist for producing Ag–AgCl electrodes. A simple method for chloridizing a silver electrode is the following: An electrolytic cell is established as shown in Fig. 6.14. The electrode, which should be clean, high-purity silver, is made the anode. The choice of electrolyte is not critical, and both dilute NaCl or HCl are frequently used. The cathode material is not particularly critical, but it is recommended that platinum wire or a platinum plate be used for the cathode. The concentration of the electrolyte is not critical, although to insure a stable electrode, doubly distilled H_2O and reagent grade chemicals should be used in making the electrolyte. The silver anode should be cleaned in *aqua regia* and washed with distilled H_2O. Recommended electrolytes are HCl in concentrations from 0.05 to 1.0 N, KCl at 0.05N, and NaCl at 0.05–0.15N. Current in the cell (Fig. 6.14) is limited to about 1 mA/cm^2 of anode surface by adjusting R. If $V = 1.5$V DC, then R will be on the order of 1 $\sim$ 10 kΩ depending upon anode size. The length of time that chloridizing is carried out determines the depth of the AgCl layer. Generally from 10 to 25% of the silver core should be converted to AgCl for stable electrodes. For small electrodes and wires, about 5 min will generally suffice. In any given situation, chloridizing time will have to be determined experimentally.

Since AgCl is photoreactive, the color of the finished electrode will depend upon the amount of light present during its preparation. Electrodes produced in the dark are usually dark colored, either sepia or plum shade. Ag–AgCl electrodes produced in light are generally gray, tan, or pink. There are conflicting statements in the literature concerning color and the concomitant quality of the electrode.

The potential of a silver–silver chloride reference electrode as a function of temperature (t = °C) over the range 0–95°C is (Bates and Bower, 1954):

$$\text{Potential (volts)} = 0.23659 - 4.8564 \times 10^{-4}\, t \\ - 3.4205 \times 10^{-6}\, t^2 + 5.869 \times 10^{-9}\, t^3$$

There are several considerations in the use of Ag–AgCl electrodes. In making the electrical connection to the electrode one must be quite careful. If solder is used, it should be coated with a waterproof insulating material to prevent its coming into contact with the electrolyte in which the electrode is immersed. This is also true of the connecting wire. Otherwise contamination of the electrode may occur because of chemical reactions between the solder or wire and the electrolyte.

AgCl is photosensitive (to ultraviolet light) and is so decomposed. It also produces a photovoltaic potential. Generally, Ag–AgCl electrodes should be stored in the dark and either used in subdued light or protected from light while in use. Excessive electrical noise produced by a given electrode may be indicative of light damage.

Ag–AgCl electrodes require Cl^- ions for proper operation. When used in biological electrolytes, they have a sufficient supply of Cl^- ions available. If they are used as skin-surface electrodes in such applications as EEG or ECG recording, it is necessary to use a wetting solution or paste that contains Cl^- ions.

Ag–AgCl electrodes are current limited because they are reversible electrodes. Sustained passage of high direct currents results in either conversion of the electrode to pure silver (if used as a cathode) or conversion of all of the silver to AgCl (if used as an anode). Generally, these electrodes are used for signal recording, as opposed to stimulating electrodes, and operate into a high-input-impedance recording circuit. Electrode current is usually $< 10^{-9}$ A.

Because Ag–AgCl electrodes are thermodynamically reversible, they exhibit (after stabilization) low noise and theoretically zero electrode-polarization impedance effects. They do produce a steady electrode potential, however, which in turn produces a DC offset in direct-coupled systems, which may require compensation. This situation may cause problems in the sensing of low-level DC potentials. Because of the electrochemical nature of these electrodes, each one assumes an absolute potential. When two such electrodes are used as a sensing pair, a DC potential difference exists (frequently of the order of a few millivolts). If this potential difference remains constant, any measurement of a bioelectric potential is unaffected, except for a steady baseline elevation. In the usual case, however, the resting potentials of the two electrodes change unequally with time and environmental temperature. This results in objectionable baseline drift in experimental determinations.

Recently a new, and reportedly much more stable, solid silver–silver chloride pellet electrode has been introduced by Beckman Instruments. These do not appear to exhibit as seriously the undesirable qualities of their fluid-bathed progenitors.

6.5.3.   The Calomel Electrode

There are a number of designs for the calomel half-cell, which uses mercury and calomel (mercurous chloride) as the reactants. Later in this chapter, the mercury–mercurous sulfate electrode, used in standard emf cells, will be discussed. Calomel electrodes are used as reference half-cells, especially in pH determinations.

The basic electrode is shown in Fig. 6.15. Since calomel is relatively insoluble, KCl solution is used as the electrolyte. The electrochemical reactions that describe the half-cell are rather complex and are omitted here. The emf of the calomel electrode depends upon the concentration of the KCl solution. The table shown below indicates the range of values which can be expected.

Calomel Half-Cell emf[a]
at 25°C

KCl	emf, V
Saturated	0.24
$1N$ KCl	0.28
$0.1N$ KCl	0.33

[a] International volts.

As with all reference half-cells, there are special considerations in fabrication to insure stability and reproducibility of the end product. The mercury used must be specially purified and the KCl and calomel

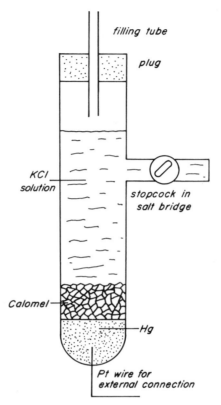

FIG. 6.15   Construction of a calomel reference electrode.

as free from impurities as possible. The calomel electrode is relatively easy to fabricate, and once constructed, is stable over long time periods. Special details in the design and preparation of these electrodes will be found in the literature.

6.5.4.   Salt Bridges

There are a number of instances in which electrical connection is necessary between parts of an electrode, between half-cells in a cell, between an electrode and a physiological system, and between an electrode and an electrolyte in which direct metallic connection is not permissible. In these cases, salt bridges are used. Let us examine several cases in which such salt bridges have been used in electrodes previously described.

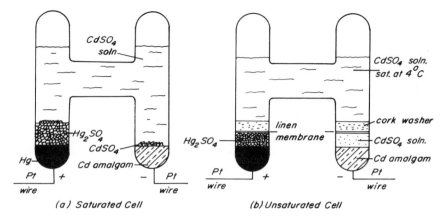

FIG. 6.16   Standard emf reference cells.

The electrolyte filled micropipet is a typical example of a salt bridge. There are cases in which it is not desirable to have direct metallic contact to a physiological system because of electrical considerations, toxicity, and trauma. In the micropipet a fluid coupling agent (salt bridge) provides the transition between the living system and the metallic electrode connection to associated electronics. Salt bridges are also used to reduce AC electrode polarization impedance effects in four-electrode systems. This application occurs incidentally in the glass micropipet. The pipet tip provides the small electrode contact area for compatibility with a physiological system, although the metal–electrolyte interface surface can be made large (thus reducing current density at the interface) by using a relatively large metal electrode in the stem end of the pipet.

Salt bridges are useful for interconnecting two half-cells to form a reference cell as shown in Fig. 6.16.

Although salt bridges are usually fluid electrolytes, such as aqueous solutions of KCl, NaCl, or $CdSO_4$, sometimes a gelatinous salt bridge is used. The gel should not be easily soluble in water at normal laboratory temperatures (20–30°C). A recipe for a suitable gel salt bridge was given by Strong (1968):

Soak 4 g of agar in 100 ml of distilled H_2O for 8–12 h. Heat the agar mixture in a beaker using a boiling water bath, not a direct flame. Heat until the agar has dissolved. Dissolve 30 g of KCl in agar solution. If necessary, add just enough distilled H_2O to effect complete dissolution of the KCl.

While the above mixture is still hot, it is used to fill clean glass U-tubes. Care must be taken to fill the tubes entirely. Air bubbles must be excluded.

Agar salt bridges have been used for making contact to simple calomel half-cells in pH determinations when platinum (quinhydrone) or antimony electrodes are used. Glass and other membranes, as discussed in subsequent portions of this book, are now available for pH measurements, and gel salt bridges are generally not used. The agar gel is subject to both chemical and bacterial contamination and must be renewed periodically. Agar salt bridges must be stored in contact with water or an electrolyte to prevent drying out. Impurities in the agar may influence sensitive measurements.

Other types of salt bridges consist of electrolyte-saturated linen wicks or bundles of camel's hair. These are useful when one wishes to minimize the trauma to delicate tissues that occurs with electrode contact. Occasionally linen membranes are used in half-cells to maintain the physical position of the reactants, especially in standard emf cell construction.

6.5.5. Reference Potential Cells—Standard Cells

In any potentiometric measurement such as pH determinations, it is necessary to have a reference standard potential against which an unknown potential voltage can be compared. Several types of potential reference cells are used. Two common ones are presented in Fig. 6.16. One configuration consists of a mercury–mercurous sulfate half-cell connected by a cadmium sulfate salt bridge to a cadmium amalgam–cadmium sulfate half-cell. This type of reference standard cell is called the saturated (normal) cell as the $CdSO_4$ electrolyte contains crystals of the salt and is a saturated solution at room temperature [Fig. 6.16(a)].

The emf produced as a function of temperature is given by the empirical relation:

$$E_s(t) = E_{s20} - 0.0000406(t - 20) - 0.00000095(t - 20)^2 + 0.00000001(t - 20)^3$$

where E_{s20} is the emf generated at 20°C, which is 1.01830 international or 1.01864 absolute volts.

Normal cells in the configuration shown in Fig. 6.16(a) are used as laboratory standards only. They are temperature sensitive and exhibit voltage hysteresis effects when heated and cooled. They are subject to mechanical shock and cannot, therefore, be transported easily.

The practical standard cell is the unsaturated type shown in Fig. 6.16(b). The $CdSO_4$ aqueous solution is adjusted to be saturated at 4°C and is thus unsaturated at usual laboratory temperatures. Unsaturated cells have the advantage that they can be transported and built in small enough sizes that they can be incorporated into electronic equipment that uses potentiometric circuits. They should not be handled roughly, however, and should be allowed to stabilize if subjected to mechanical shock or vibration.

Newly produced unsaturated cells generally yield an emf of 1.0190–1.0194 absolute volts. They age more rapidly than normal (saturated) cells and the emf can be expected to decrease by about 30 μV/yr (0.003%/yr).

There are certain precautions when using unsaturated standard cells. The permissible temperature range is 4°C $< t <$ 40°C. Standard cells should be protected against thermal shock and all parts of the cell should be in thermal equilibrium. Currents in excess of 10 μA should not be drawn from standard cells. Current should be drawn only long enough to check galvanometer deflection in potentiometric circuits. Prolonged low-current drain, or short-term high-current drain destroys standard cells. They are basically potential reference cells and are not capable of delivering power.

Except for the most exact measurements, Zener diodes in a constant temperature oven are generally used now as reference voltage sources in electrical equipment. They are stable and virtually immune to mechanical trauma.

6.5.6. Potentiometer for Voltage Measurements

The potentiometer, used for accurate determinations of voltage in standards laboratories, has various applications in instrumentation systems. One, of course, is the accurate determination of various bioelectric and biochemical potentials. In addition, many instruments such as pH meters are designed about a potentiometer circuit as a central core. The circuit consists of two batteries (or very stable potential sources). One is a working battery or voltage, and the other is a standard reference cell or zener diode. The heart of the circuit is a very accurately calibrated resistance called a precision slide wire (PSW). It is usually in the form of a helix wrapped about a solid core. Such a circuit appears in the pH meter schematic shown in Fig. 6.17. With reference to this figure, operation of a potentiometer is as follows: Switch 3 is connected to the "cal." position and p.b. switch 2 is

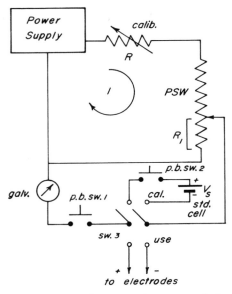

FIG. 6.17   Circuit schematic for potentiometer and simple pH meter.

depressed. P.b. switch 1 is depressed at intervals while the PSW sliding contact is adjusted. When the voltage drop across the section of the PSW designated as R_1 is equal to V_s, no deflection of the galvanometer will be noted. Since a zero-current balance method is used, the internal resistance of the galvanometer can be ignored. The balance value of R_1 is now recorded. The standard cell cannot be permanently connected, as under unbalance conditions, current exists in the galvanometer circuit, and this would eventually damage the standard cell. P.b. switch 2 should now be deactivated. Switch 3 is now connected to the "use" position. The PSW is again adjusted for balance, while intermittently depressing p.b. switch 1. The new balance resistance value R_2 is recorded. The defining equations are:

$$V_s = R_1I \qquad V_u = R_2I$$

The current I must remain constant during the measurement and V_u is the unknown voltage to be determined. The value of I is initially set at some convenient level by adjusting the calibration resistance R. The power supply voltage must also remain constant during the measurement. Combining the two equations:

$$\frac{V_u}{V_s} = \frac{R_2}{R_1} \qquad \text{and} \qquad V_u = V_s\frac{R_2}{R_1}$$

Thus V_u is determined when the ratio (R_2/R_1) and the standard cell emf V_s are known.

A direct reading potentiometer can be made by normalizing R_1 to 1 Ω and V_s to 1 V, then V_u (volts) $= R_2$ (ohms calibrated as volts). The PSW is calibrated directly in voltage units (or pH units in the pH meter). The rheostat R can be used to zero-adjust the potentiometer for the required normalization. Many commercially available potentiometers are direct reading, with built-in provision to adjust for different standard cell potentials. A typical value for V_s is 1.019 V abs. for an unsaturated cell.

The use of the potentiometer circuit in pH meters is discussed briefly in the next section.

6.6. pH ELECTRODES AND pH METERS

A number of electrode configurations have been proposed for the measurement of pH. We will discuss in detail only one of them here. The electrodes that are the most easily fabricated are generally the poorest for pH measurements. These are the antimony and the quinhydrone electrodes.

6.6.1. The Glass Electrode

The electrode most commonly used in pH determinations is the glass membrane electrode shown in two configurations in Fig. 6.18(a) and (b). In operation, the glass electrode depends upon phase boundary potentials that form between the glass surface and the external electrolyte. Glass surfaces also acquire charge by ionic absorption, although it is thought that this occurs to any degree only with excellent dielectrics such as quartz.

In current practice, the reference electrode is either calomel or silver–silver chloride and a glass electrode is the active element. The electrodes and pH meter potentiometric circuit are calibrated against standard reference solutions. Under the assumption that liquid junction potential does not change if unknown solution U_1 is replaced by U_2, the Nernst relation would be

$$\Delta E = \frac{2.3026RT}{F} \Delta(-\log \alpha_H)$$

where $-\log \alpha_H$ is the pH and ΔE is the change in potential when U_1 is replaced by U_2.

If we now let one of the U terms represent a standard reference solution, then we can substitute explicitly in the Nernst relation

$$E_u - E_{\text{ref}} = \frac{2.3026RT}{F} \left(-\log \alpha_{H,u} + \log \alpha_{H,\text{ref}} \right)$$

$$pH_u = pH_{\text{ref}} + \frac{(E_u - E_{\text{ref}})F}{2.3026RT}$$

Glass electrodes are not quite linear and the pH response of these electrodes depends strongly upon the type of glass that forms the bulb [Fig. 6.18(a)] or the membrane [Fig. 6.18(b)]. Glass electrodes usually deviate from true pH values in the alkaline region. This is related to attack by strong alkalis upon the electrode and concomitant chemical breakdown of the glass. Alkaline pH errors appear to be ion specific and are related to sodium ions. In an attempt to combat this situation, lithium glass rather than soda lime glass has been suggested. Errors in

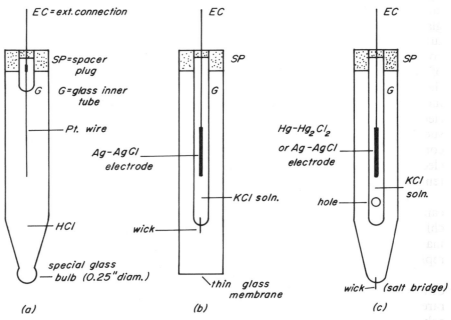

FIG. 6.18   (a) and (b) Glass electrodes for pH measurements; (c) reference electrode.

pH reading also occur in the acid range for very low pH values. Various factors appear to be causes for reading deviations, but they appear mainly to be related to glass formulae.

Glass is hygroscopic and this generally leads to electrode break-down after nine months to two years. Water tends to leach out alkali constituents in the glass, and the resulting lye breaks down and cor-rodes the membranes. On the other hand, the electrode must be hydrated in order to yield accurate pH readings. If glass electrodes are stored dry, they must be immersed in aqueous solution and allowed to stabilize before use. Generally storage in distilled water is preferable and one must accept the fact that pH electrodes do not have long lives. Electrode life is directly related to membrane thickness, which ordinarily ranges from 50 to 140 μm. Deposits should not be permitted to form on the membrane since they are difficult to remove. Mechanical cleaning is impossible because of membrane fragility.

The reference electrode used in conjunction with a pH glass electrode is generally of the salt-bridge type and a typical configuration is shown in Fig. 6.18. Use of a salt bridge (wick type in this case) protects the reference electrode from chemical action and contamination by the test solution. The electrode consists of an inner glass tube that contains mercury and calomel, and is in turn surrounded by an outer glass tube filled with saturated KCl solution. A small hole in the inner tube permits contact of the KCl solution with the mercury–calomel combination. Contact with the sample is usually established by either of two methods. In one case, a small fiber is sealed permanently into the immersion end of the tube. In the other, KCl reaches the sample by means of a small hole in the immersion end beneath a ground glass sleeve. A modification uses a ground glass plug in the immersion end, such that KCl solution seeps through the plug–tube juncture and contacts with the sample. The calomel electrode is a practical reference electrode and is very stable with respect to potential, provided that temperature is reasonably controlled.

In some cases, the solution for which pH is to be determined cannot tolerate the calomel electrode. In such cases, a silver–silver chloride electrode is used. The electrode is constructed in the same manner as the calomel electrode, except that silver–silver chloride replaces mercury–calomel in the center tube.

There is a problem however, with Ag–AgCl reference electrodes. Silver in the ionized state does not achieve the electronic structure of a rare gas. Because of this situation, when used in high-protein-containing solutions such as blood, Ag–AgCl electrodes become poisoned by adhesion of material to the electrode surface.

6.6.2.   pH Meter

A schematic diagram for a simple pH meter is shown in Fig. 6.17. It is nothing more than a conventional potentiometer circuit. The precision slide wire (PSW) is calibrated in units of pH rather than in volts or ohms. A stable power supply is used, which may be a battery or a regulated electronic supply. The reference voltage source may be an unsaturated standard cell or a Zener diode electronic reference.

In use, initial calibration of the meter is accomplished by setting the PSW to the calibration point (usually pH 7.0), the switch to CAL, and adjusting R for null indication of the meter. The switch is then set to USE and pH is determined by adjusting the PSW until the meter nulls. Should the PSW have to be recalibrated (different pH electrodes, etc.), buffered solutions of known pH are used.

The circuit shown in Fig. 6.19 represents an electronic version of Fig. 6.17. Electronic power supplies are used and an electronic voltage comparator with meter replaces the galvanometer.

Many degrees of sophistication exist in modern pH meters. Some are direct reading and operate on the basis of a voltage offset against a standard voltage reference in the instrument.

6.6.3.   Ion-Specific Electrodes

Most ion-specific electrodes operate on a potentiometric principle rather than an amperometric or polarographic principle; that is, a change in potential is sensed rather than a change in current.

The number and variety of ion-specific electrodes is rapidly increasing with no end in sight. At the present writing, it is possible to use such electrodes to determine, either by direct or indirect measurement, ionic concentrations of the following species: ammonia, bromide, cadmium, calcium, chloride, cupric, cyanide, fluoride, fluoroborate, iodide, lead, nitrate, perchlorate, potassium, sulfide, sodium, sulfur dioxide, thiocyanate, by direct measurement, and by titration methods: aluminum, boron, chromium, cobalt, magnesium, mercury, nickel, phosphate, silver, sulfate, zinc. Historically the glass pH electrode was the first of the ion-specific membrane electrodes.

Although many membrane materials are available, glass is frequently used in biological work. Other membrane materials are discussed briefly in this chapter. Many sorts of membranes have been developed for testing and monitoring use in chemical manufacturing processes and are not directly applicable to physiological measurements. Many membranes function satisfactorily when only one ionic

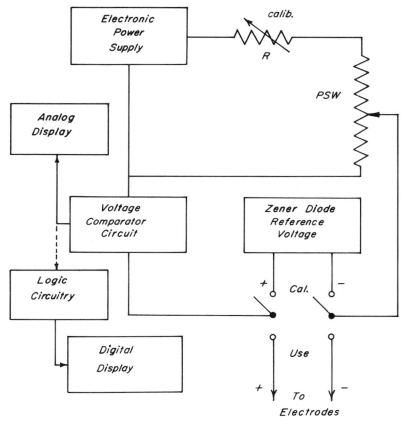

FIG. 6.19   System schematic for electronic pH meter. An optional connection to a digital display is shown.

species is present. Biological electrolytes tend to be polyionic, which limits, to some extent, the types of membranes that can be used. In single ionic systems, or where the differences in ion characteristics are pronounced, membrane electrodes can provide both a reliable and effective system for measuring ion concentration. Concentration is not measured directly, but rather is inferred from a measurement of ion activity.

In biological and environmental studies, we are normally interested in cationic determinations. The membrane-electrode potentials which are observed are described by the Nernst relation:

$$E_{\text{obs}} = E° - \frac{RT}{nF} \ln (\alpha_c{}^+)$$

where E_{obs} is the potential observed for a given activity of cation c^+ (V), $E°$ is the potential of ion being determined [when the energy of the cation is equal to the standard state, $E_{obs} = 0$ (volts)], R is the universal gas constant, T is the temperature in deg K, n is the number of electrons transferred in the reversible reactions, F is Faraday's constant (96,493 C/mol), and α_c^+ is the activity of the cation under study.

The membrane serves as a phase separator that is made selectively permeable to those ions for which a concentration value (activity) is to be determined. Using the Nernst relation, with necessary correction factors for temperature and nonideality of the membrane, the potential between the known and unknown concentration can be measured and the unknown ionic concentration thus determined.

A number of materials have been developed for use as electrode membranes. These include various glass formulae, plastics, ceramic and clay materials, collodion matrix membranes, and heterogeneous membranes composed of ion-exchange resins imbedded in an inert binder. Additional techniques include liquid ion-exchange membranes and mixed crystal membranes. Typical electrodes are shown in Fig. 6.20.

There are various problems that arise in the use of ion-selective electrodes. Specific electrodes measure activities rather than concentrations. Analytical methods for making the required conversion are

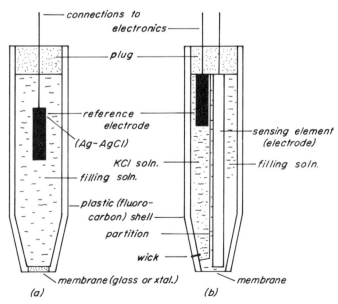

FIG. 6.20   Simple (a) and combination (b) membrane ion-specific electrodes.

available. For commercial electrodes, the manufacturer can supply the necessary information.

Some ionic species will interfere with other ionic species thus distorting the voltage reading produced by the electrode. Zinc interferes with the calcium electrode. Hydroxide interferes with the fluoride electrode. Sugar solutions produce low readings from pH and calcium electrodes that have been calibrated in aqueous standard solutions.

To avoid problems of this nature, solutions should be prepared very carefully to preclude inclusion of interference ions, or buffered to adjust pH. Calibration should be carried out in solutions of the same general chemical composition as the solutions in which measurements are to be performed. Again, manufacturer's specification sheets should be consulted for information on particular electrodes.

Decomplexing agents can be added to some samples to render inert those ionic species that either interfere directly with electrode operation, or complex with the ion being measured.

Ion-selective electrodes may be either of dipping or flow-through design. The former are used for general purpose, whereas the latter find specific application in anaerobic measurements.

6.6.3.1. *Calibration Curves*

Ion-specific electrodes should be calibrated periodically to insure that the electrodes are stable. This is usually accomplished by measuring the electrode potential in standardizing solutions produced by serial dilution. A calibration curve is then prepared using semilogarithmic paper. Electrode potential is plotted on the vertical linear scale, whereas concentration is plotted on the horizontal logarithmic scale. If one desires to obtain concentration values for unknown solutions by direct measurement, then the ionic strengths of the standardizing solutions and the unknown samples must be similar.

Ionic strength adjustors can be added in equal amounts to both the standardizing and unknown solutions to damp out ionic strength differences. Generally a high level of a noninterfering electrolyte is added to produce a high but constant ionic strength in both classes of solutions.

pH adjustments may also be necessary for several reasons. Strongly acidic or basic samples can damage sensitive membranes. In addition, measurements of cyanide species must be done in basic solutions to prevent generation of toxic HCN gas.

Ion-specific electrodes are used in conjunction with reference electrodes. Frequently a pH reference electrode, such as a conventional calomel reference, should not be used, since KCl may not be a suitable electrode filling agent for many samples and in particular, for measurements of the silver ion. A reference electrode that is designed specifically for ion-selective measurements should be used. (In some cases ammonium nitrate may be substituted as the filling electrolyte.)

6.6.3.2. *Use Extension of Existing Electrodes*

In some cases ionic species for which no electrode exists can be measured using an electrode which is specific for another ionic species. Basically an indirect measurement is made that involves some chemical reaction between the ionic species of interest and the ionic species to which the electrode is selective.

Titration techniques may be applied so that a fluoride-selective electrode can be used to measure aluminum ionic strength.

An iodide-selective electrode can be used to determine dissolved chlorine. Iodide is added when the sample solution is prepared. Loss of iodide (by reaction with chlorine) is then measured by the electrode.

6.6.4. The Glass Membrane Electrode

Practical ion-selective electrodes for analytical and clinical use generally use glass membranes. Glasses are used that are similar in composition to pH glass. This means that, in use, glass electrodes for specific ion measurements must be corrected for pH. The technique for doing this is presented below. The majority of the ion-selective electrodes available for biological work are cationic-selective and respond to sodium, potassium, calcium, lithium, or ammonium ions. Electrodes are also available that respond to the ions of cadmium, copper, lead, and silver. Anionic-sensitive electrodes include those sensitive to Br^-, Cl^-, CN^-, F^-, BF_4^-, I^-, NO_3^-, and ClO_4^-.

The glass membrane electrode is described by the half-cell:

Glass membrane	electrolyte	Ag–AgCl electrode	$\longrightarrow$	connection to potentiometer

The filling electrolyte varies in composition and is subsequently different for specific electrodes.

6.6.4.1. *Calibration*

As we have pointed out previously, electrodes respond to ion activity rather than ion concentration. In addition, there is lack of linearity between theoretical (from the Nernst relation) and measured values of emf for a given spread of ion activities for a particular ion in aqueous solution.

In the laboratory, a calibration curve is developed for each ion-selective electrode and the calibration is checked periodically to detect membrane degradation. From the Nernst relation, we can compute that the potential at the glass membrane changes on the order of 59 mV for each change by a factor of ten in ion concentration at 25°C, for monovalent ions.

To avoid the problem of activity versus concentration, a direct calibration of the electrode is carried out in known electrolytes. Typically one prepares known solutions by dissolving known amounts of a given salt in doubly distilled or deionized water. Commercial electrodes frequently respond over the range 0.0001–1.0N. It is usually satisfactory to adjust the concentration steps to 0.01N, 0.03N, 0.1N, etc. For each concentration, the electrode emf is measured against a reference electrode (Ag–AgCl or calomel) at constant temperature. The electrodes must be carefully rinsed with distilled water between samples. In this manner, an emf versus moles per liter calibration curve can be plotted, as shown in Fig. 6.21. For increased accuracy over a given range of concentrations, additional data points can be obtained by using additional reference solutions. In calibrating an ion-selective electrode and in using it later on, one must be sure that the entire active membrane surface is completely immersed in the test solution.

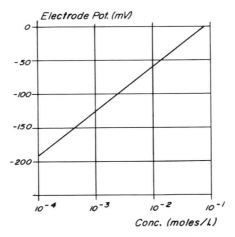

FIG. 6.21   Typical working calibration curve for glass membrane sodium ion-selective electrode.

6.6.4.2.   *Correction for pH*

Cationic-selective electrodes generally respond to hydrogen ions as well as the ion of interest. Thus one must adjust a test solution to suppress the influence of the hydrogen ions. This is achieved by adjusting the hydrogen ion concentration so that it is at least four orders of magnitude lower than the lowest ion concentration to be measured.

As an example, let us suppose that sodium ion concentration is to be measured over the range 0.0001–$0.1 M$.

$$pNa = -\log Na^+$$

For $0.0001 M$ sodium ion concentration

$$pNa = -\log 10^{-4} = 4$$

For $0.1 M$ sodium ion concentration

$$pNa = -\log 10^{-1} = 1$$

Thus, if the lowest sodium ion concentration is $pNa = 4$, or 10^{-4}, the hydrogen ion concentration must be suppressed to 10^{-8} or $pH = 8$.

This can be accomplished by adding a basic salt of different composition from the test salt. For example, if sodium is the test salt, calcium hydroxide could be added. One must be careful to use analytical-grade reagents to avoid adding more of the test element to the sample. For convenience, one can simply saturate the test solution with the basic salt to assure pH suppression. Normally the test solution is modified for pH in the range 8–12. Some glass membranes are damaged by strong acids and alkalis as mentioned previously.

Care of ion-selective glass membrane electrodes is similar to that suggested for glass pH electrodes. One must be careful to maintain the correct amount of filling electrolyte.

Glass membrane electrodes have a limited life, usually 1–2 yr if cared for. The glass becomes slowly weakened by contact with sample and standardizing solutions until it decomposes or shatters.

Some glass membranes that are selective for the potassium ion are also highly sensitive to acetylcholine. Thus electrodes made from this type of glass should not be used in studies where acetylcholine is present (e.g., synapse studies in neurophysiological research).

Membrane electrodes, although they can be made to measure concentrations in the case of single ions, are rarely ion specific in the more general polyionic situations.

Only when one ion species is highly hydrated, or otherwise large compared with other ions, or very insoluble in the membrane, will the membrane show a strong specific ion selectivity. Selection between ions located near one another in the periodic table is seldom very good.

6.7.   THE OXYGEN ELECTRODE—
A POLAROGRAPHIC ELECTRODE

With the development of plastic membranes that are selectively permeable to the molecules of oxygen and carbon dioxide, but not to ions and water molecules, it has become possible to construct specific electrodes for pO_2 and pCO_2 measurements.

The basis for the operation of an oxygen electrode is that oxygen gas in solution reacts with a negatively charged (polarized) metal surface and forms OH^- radicals. Proteins, in particular, and other substances in solution, however, are also attracted to such surfaces and produce "poisoning" or reduced sensitivity to oxygen. In 1956, Clark developed a membrane electrode in which he used a polyethylene membrane to isolate the test solution from the metal portions of the electrode. The basic electrode configuration is shown in Fig. 6.22. A platinum cathode is used and is formed from 10 to 25 μm diameter platinum wire. Early electrodes used 2 mm diameter wire, but it has been found that a smaller cathode requires less oxygen for operation and produces a smaller pO_2 gradient in the vicinity of the electrode; this matter will be taken up in greater detail subsequently.

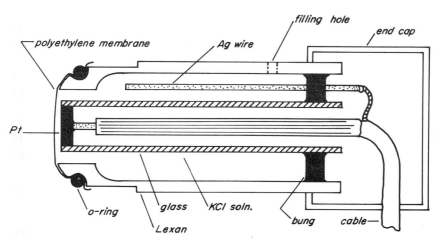

FIG. 6.22   The basic Clark pO_2 electrode.

Normally the silver electrode is silver wire with a diameter usually larger than that of the platinum. The silver wire is actually a silver–silver chloride reference electrode and is chloridized before the pO_2 electrode is used.

Electrode response time is relatively slow since one must wait for oxygen to diffuse through the membrane. The basic reaction at the cathode is probably

$$\tfrac{1}{2}O_2 + 2H^+ \text{ (in solution)} = HOH - 2e^-$$

It is not unusual for the initial response time for an oxygen electrode to be as long as 1 min. In general, the smaller the cathode tip, the faster the response. The membrane affects response time through its permeability to oxygen. Typical membranes are PVC, Mylar, cellulose acetate, polyethylene, polypropylene, silicone rubber, and ethyl cellulose.

For blood oxygen measurements, Eastman polypropylene has been found the most satisfactory material to use.

Oxygen electrodes do exhibit aging effects and these have several causes. The manifestations of aging are increased polarization current, increased response time, and sensitivity to hydrostatic pressure. Damage to the membrane or membrane aging is frequently a problem and is easily rectified by replacing the membrane. Another cause of aging is deposition of silver around the edges of the platinum cathode. This problem may be solved by carefully cleaning the electrode tip on wet Arkansas stone; emery should not be used since it may imbed in the electrode surface.

The oxygen electrode is polarographic and in operation requires a polarizing potential of from 0.2 to 0.9 V, with the platinum electrode negative relative to the Ag–AgCl wire. A circuit schematic is shown in Fig. 6.23. Current through the cell is then a linear function of the oxygen tension in the solution that bathes the two electrodes. When the platinum is made slightly negative relative to the silver, the oxygen reaching the platinum is reduced electrolytically. The reaction is not completely understood, although a probable partial description has been given above. Normally, about 0.2 V potential difference is sufficient. When the platinum is made more negative (-0.6 to -0.9 V) relative to the silver, the reaction rate of the electrolytic reduction is limited by the maximum rate at which oxygen can diffuse through the membrane to the electrode surface. At this point, the magnitude of the potential difference between the electrodes has little effect upon output current, and the output current variation is directly proportional to oxygen concentration in the bathing solution.

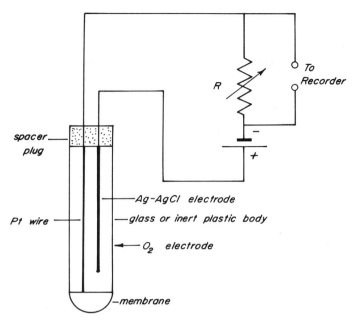

FIG. 6.23   Simple polarographic circuit for pO_2 measurements.

To decrease response time and prevent erroneous readings because of oxygen concentration gradients near the electrode, the test solution should be stirred, if possible. In this way, a homogeneous sample is assured. As is the case with other membrane electrodes, the oxygen electrode is pH sensitive. It is thought that at low pH levels, H_2O_2 may be formed rather than OH^-, thus using only half as many electrons per mole of oxygen. Oxygen permeable membranes are also permeable to CO_2, and this gas freely diffusing into the solution bathing the platinum and silver–silver chloride electrodes alters the pH. The pH effect of CO_2 can be eliminated by stabilizing the test solution either at pH 7, by using a pH 7 buffer stock solution to which $0.1M$ KCl has been added, or at pH 9, by using $0.5M$ $NaHCO_3$ and $0.1M$ KCl. It has been reported that the pH 9 buffer best minimizes the CO_2 effect.

Various oxygen electrode configurations have been developed, including a miniaturized needle electrode for tissue and arterial use.

6.8. THE CO_2 ELECTRODE

Figure 6.24 illustrates how an electrode for CO_2 measurement might be constructed. It is essentially a glass pH electrode, bathed in an

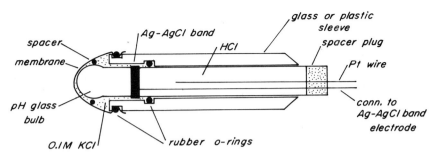

FIG . 6 24   Electrode for pCO_2 measurements.

electrolyte, and covered by a membrane freely permeable to CO_2. A silver–silver chloride reference electrode is incorporated into the system to form a full cell for direct pH measurement. In use, the electrode is dipped in the test solution and the output voltage read as pH on a pH meter. Since the electrode responds to the absolute value of pCO_2, it does not differentiate between gas or liquids of the same pCO_2 at constant temperature. Thus the electrode can be calibrated by flowing known concentrations of CO_2 in nitrogen or air over the electrode and calibrating the pH units on the pH meter in terms of pCO_2.

Various membrane materials can be used including rubber, silastic, Teflon, and polyethylene. Membrane thickness determines electrode response time with thin membranes (0.001 in.) being the most responsive. Response time is similar to that of pO_2 electrodes (1–2 min), which is to be expected since the same sorts of processes are involved.

The electrical output from the pCO_2 electrode is about 58–59 mV per 10:1 change in pCO_2 at 37°C (body temperature).

The most satisfactory filling electrolyte in terms of linearity of response and agreement between theoretical and experimental electrode behavior is $0.01M$ $NaHCO_3$ (or $KHCO_3$).

Measurement errors in the use of this electrode relate to membrane leaks (pin holes, etc.), inadequate temperature regulation during measurement, insufficient sample size, and air bubbles (in liquid samples) adhering to the outside of the membrane.

Combination electrode assemblies (pO_2–pCO_2) have been designed for clinical use in making blood gas measurements.

6.9. OTHER TECHNIQUES

A number of additional ion-specific electrode techniques have been developed. These include use of enzymes and antibiotics bonded into a membrane matrix. Ion-specific microelectrodes have been developed that use ion-exchange resins.

The aerospace program has brought about considerable research on electrodes for long-term monitoring. These include self-wetting and spray-on electrodes, as well as various clip-on and "safety pin" electrodes.

6.10. REFERENCES

Adams, R. N., 1969, *Electrochemistry at Solid Electrodes*, Dekker, New York.

Bates, R. G., and V. E. Bower, 1954, *J. Res. Nat. Bur. Std.* **53**, 283.

Berman, H. J., and N. C. Hebert, eds., 1974, *Ion-selective Microelectrodes*, Plenum, New York.

Ferris, C. D., 1974, *Introduction to Bioelectrodes*, Plenum, New York.

Geddes, L. A., 1972, *Electrodes and the Measurement of Bioelectric Events*, Wiley-Interscience, New York.

Ives, D. J. G., and G. J. Janz, 1961, *Reference Electrodes*, Academic Press, New York.

Lakshminarayanaiah, N., 1976, *Membrane Electrodes*, Academic Press, New York.

Miller, H. A., and D. C. Harrison, eds., 1974, *Biomedical Electrode Technology: Theory and Practice*, Academic Press, New York.

Newman, J., 1972, *Electrochemical Systems*, Prentice-Hall, Englewood Cliffs, N.J.

Strong, C. L., 1968, *Sci. Amer.* **219** (3), 232.

Section 3

AMPLIFIERS AND SIGNAL CONDITIONING

7

Preamplifiers for use with Transducers and Bioelectrodes

The purpose of this chapter is not to attempt a detailed presentation of circuit design techniques for preamplifiers, but rather to discuss certain important considerations in their design and use. With micro-circuits and chips now becoming readily available at low cost, it will soon be unnecessary to design an amplifier system from discrete elements. There are, however, important considerations concerning input characteristics and dynamic response that are common to all amplifier systems. These are the topics we shall review in the present chapter.

The signal input into a preamplifier may be a direct connection (DC amplifier) or it may pass through a capacitor that blocks a direct current component in the signal (AC amplifier). Direct coupled or DC amplifiers are more difficult to use since any direct current component in the input signal tends to offset the operating point of the amplifier. Special design techniques are required to insure stable system operation. AC amplifiers, on the other hand, do not suffer from DC offset

151

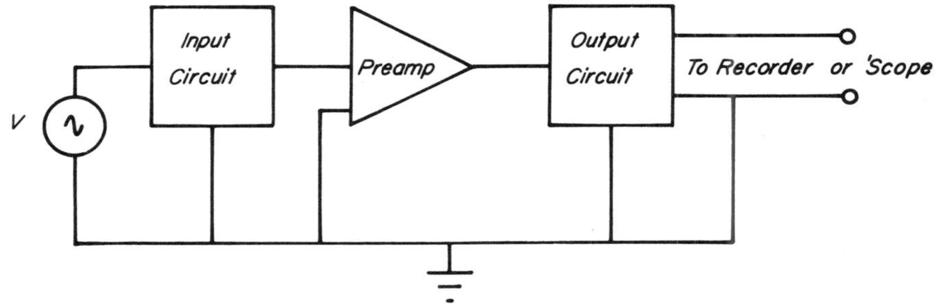

FIG. 7.1 Basic preamplifier system; V is the signal source.

problems, but because of the input DC blocking capacitor, their low frequency response is limited and signal distortion can occur. A compromise is the chopper amplifier that, in effect, converts DC to AC signals for amplification purposes, and then converts the output back to DC.

Another consideration in the use of amplifiers is the nature of the signal source with respect to a reference ground. Some signals may be processed on a single-ended basis (signal lead and ground lead, as in recording from a single cell); other signals are processed on a difference basis (two signal leads and a reference ground, as in electrocardiograph recording). In the first instance, a simple single-ended input circuit preamplifier suffices; in the second instance, a differential amplifier is required. These are some of the matters considered in this chapter.

The basic system under consideration is illustrated in Fig. 7.1. The system shown is a simple single-ended preamplifier; differential amplifiers will be considered as a separate topic.

7.1. PREAMPLIFIER INPUT CONSIDERATIONS

The basic input circuit for a general preamplifier is shown in Fig. 7.2. A triangle is used to represent the amplifier and associated power supplies that provide the necessary bias voltages for its operation. In the case of a DC amplifier, C_c is replaced by a short circuit.

In the circuit, C_c represents the AC coupling (DC blocking) capacitor; C_i is the combination of shunt stray capacitance (from input cables, etc., to ground) and the dynamic input capacitance of the amplifier itself; and R_i is the input resistance of the amplifier system,

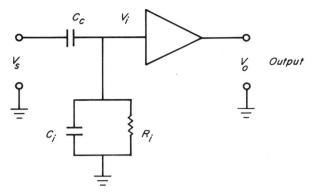

FIG. 7.2   Generalized input circuit; V_s is the signal input.

which combines input bias resistance and dynamic input resistance of the active element in the amplifier.

The acutal signal V_i which reaches the active circuitry of the preamplifier is modified from V_s by the input circuitry and is given by (Laplace transform notation):

$$V_i = \left(\frac{sR_iC_c}{sR_i(C_i + C_c) + 1}\right)V_s = \left(\frac{s\tau_1}{s\tau_2 + 1}\right)V_s$$

The input coupling circuit is effectively a high-pass filter, or it can be considered as a differentiating circuit at high frequencies with the time constant: $\tau_2 = R_i(C_i + C_c)$. The frequency response (V_i/V_s versus ω) of the input circuit is shown in Fig. 7.3.

From an examination of Fig. 7.3, we see that we can expect signal distortion because of attenuation of the low frequency components

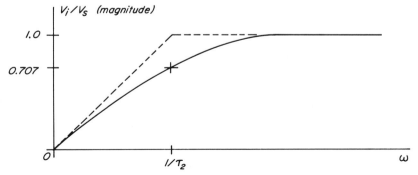

FIG. 7.3   Frequency response of input circuit; $\tau_2 = R_i(C_i + C_c)$.

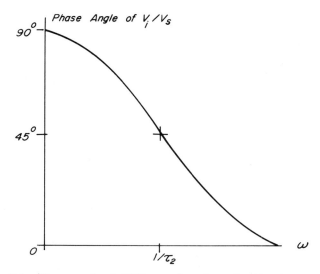

FIG. 7.4   Phase angle of V_i/V_s as a function of radian frequency.

in the signal. At low frequencies, the signal components are shifted in phase, which produces additional waveform distortion. Phase angle versus frequency values for the input circuit are plotted in Fig. 7.4.

When direct coupled (DC) amplifiers are used, C_c is replaced by a short circuit. In this case, theoretically, there is no signal distortion. The R_iC_i combination, however, does represent an electrical load on the signal source and at high frequencies, when the shunting effect of C_i becomes important, there is attenuation of the high frequency components. This is especially important in using micropipet electrodes, as described below.

7.1.1.   Special Considerations for Microelectrodes

A rudimentary representation for an input circuit with glass micro-electrodes (using a DC amplifier) is shown in Fig. 7.5, in which R_e represents the series resistance of the glass micropipet. A detailed description of electrode polarization problems was presented in Chapter 6. For this discussion it is sufficient to consider only the elements shown in Fig. 7.5 since any signal distortion produced by electrode polarization can be incorporated into the description of V_s.

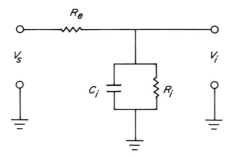

FIG. 7.5 Simple microelectrode circuit model.

The voltage V_i presented to the preamplifier is:

$$V_i = \left(\frac{R_i}{sR_iR_eC_i + R_i + R_e}\right) V_s$$

The input circuit is a combined voltage divider (attenuator for all frequencies) plus a low-pass filter with time constant $R_iR_eC_i/(R_e + R_i)$. If we neglect C_i for the moment, then

$$V_i/V_s = R_i/(R_i + R_e)$$

If we consider a typical commercial preamplifier such as is found in a high-gain unit in an oscilloscope, $R_i \sim 3 \times 10^6 \ \Omega$; R_e is usually about $10^8 \Omega$. Thus

$$V_i/V_s \sim 3 \times 10^6/10^8 = 0.03$$

With conventional amplifiers, the signal is attenuated by two orders of magnitude. Hence a signal of 100 mV at the tip of the micropipet would be only 3 mV at the input terminals of the preamplifier. Even when R_i is increased to $10^8 \ \Omega$

$$V_i/V_s = 0.5$$

and half of the signal is lost. We see, then, that the input resistance R_i of a preamplifier to be used with micropipets must be of the order of $10^9 \ \Omega$ or higher. This requires the use of an electrometer input system to be described subsequently.

Now if we consider the effect of C_i we note another problem. Typical values of C_i are from 2–5 pF. Thus for $R_i = 3 \times 10^6 \Omega$, the minimum time constant is:

$$\tau = R_iC_i = 3 \times 10^6 \times 2 \times 10^{-12} = 6 \times 10^{-6} \ s$$
$$= 6 \ \mu s, R_e \sim 0 \qquad \text{the metal microelectrode case}$$

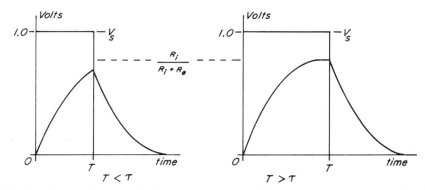

FIG . 7.6   Pulse distortion by input capacitance; T = pulse duration; τ = circuit time constant.

For glass microelectrodes, R_e is large and the time constant is:

$$\tau = R_e R_i C_i/(R_e + R_i)$$
$$= 5.82 \ \mu s, \ R_e \sim 10^8 \ \Omega \qquad R_e \sim 10^6 \ \Omega \text{ for large tip diameters}$$

When R_i is increased to $10^9 \ \Omega$, the minimum time constant is 2 ms. This means that the 3 dB cutoff frequency for the input circuit is only 80 Hz. So that if we increase R_i to prevent signal attenuation, C_i causes a substantial loss in frequency response and subsequent signal distortion. Figure 7.6 is a graphical illustration of this situation. The solution to this problem lies in using a feedback preamplifier system which reflects a negative capacitance at its input terminals, thus canceling the effect of C_i. This system is described in a subsequent section of this chapter.

The discussion above indicates why glass micropipets are generally used for low frequency recording since they are characterized by the circuit shown in Fig. 7.5. The low-pass filter aspect of Fig. 7.5 accounts for this situation. Slowly varying signals are passed without appreciable distortion or attenuation (if R_i is sufficiently large). It should be noted that most of the stray capacitances associated with micropipets (Fig. 13 of Chapter 6) can be lumped into C_i. Alternating current electrode polarization in recording electrodes sometimes can be virtually eliminated by using a silver–silver chloride connection to the electrolyte in the lumen of the micropipet. When this is done, then Fig. 7.5 is essentially a valid representation for a glass micropipet system.

When metal microelectrodes are used, Fig. 7.7 indicates an approximate input circuit configuration; R_e is nearly zero and is omitted, but C_p, the electrode polarization capacitance, now becomes

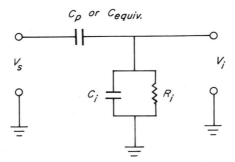

FIG. 7.7   Basic network representation for a metal microelectrode.

important. When direct coupling to the amplifier is used, only C_p is present. For AC coupling, the series capacitance is given by

$$C_{\text{equiv}} = C_p C_c / (C_p + C_c)$$

Of course, C_p is frequency dependent and its effect generally can be neglected above 1000 Hz. The circuit in Fig. 7.7 is essentially that of Fig. 7.2, and the mathematical description is similar except that C_p increases (on the series basis) as frequency increases and can be treated as a short circuit above 1000 Hz. A metal microelectrode system is basically a band-pass filter system, but with strong high-pass filter characteristics. When C_i is eliminated by using a negative-input capacitance preamplifier, then it is a high-pass filter. This is why metal microelectrodes perform best for fast pulses such as those typically found in neurophysiological processes.

We now see the reason for the general rule that glass microelectrodes are used for slowly varying processes (intercellular recording), and metal microelectrodes are used for rapid processes (neuron records).

7.2.   DYNAMIC RESPONSE OF PREAMPLIFIERS

In this section we consider such matters as frequency response, bandwidth, gain, and risetime. The basic system under examination is shown in Fig. 7.8. We represent the active portion of the amplifier by a dynamic transfer factor g_m. The input circuit is defined by a transfer function $G_i(s)$ and the output circuit by a transfer function $G_0(s)$ such that:

$$V_i = G_i(s)V_s \qquad V_0 = g_m G_0(s)V_i$$

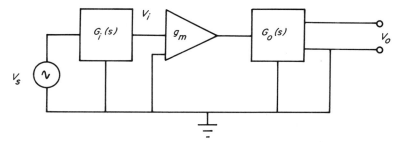

FIG. 7.8   Abstraction of a general preamplifier.

where Laplace transform notation is used. The overall relation between the system source voltage and the output voltage is then

$$V_0 = G_i(s)g_m G_0(s)V_s$$

7.2.1.   Amplifier Gain

If we designate V_i as the voltage available at the input terminals to the amplifier, then we define the gain of the amplifier through the relation

$$V_0/V_i = g_m G_0(s)$$

Since the factor $g_m G_0(s)$ can be complex if we substitute the steady-state variable $j\omega$ for s, we take the magnitude of the ratio V_0/V_i as the definition of gain. Thus

$$\text{amplifier gain} = |V_0/V_i| = |g_m G_0(s)| = K$$

and therefore

$$K = |g_m G_0(s)|$$

In the general case, if V_0/V_i is a complex quantity, then $g_m G_0(s)$ can be represented as

$$g_m G_0(s) = A + jB = \sqrt{A^2 + B^2}\ \underline{/\theta} = \sqrt{A^2 + B^2}\ e^{j\theta}$$

where $\theta = \tan^{-1}(B/A)$

The transfer characteristic of the amplifier (V_0/V_i) has now been represented in terms of a magnitude (gain) and a phase-shift angle θ, such that

$$K = \sqrt{A^2 + B^2} \qquad \theta = \tan^{-1}(B/A)$$

when the substitution $s = j\omega$ is made for the specific relations represented in A and B, we can make a gain versus frequency plot and a

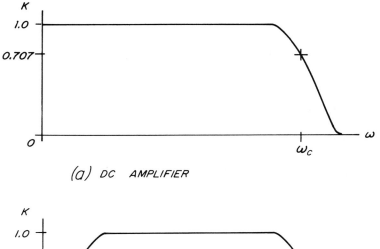

(a) DC AMPLIFIER

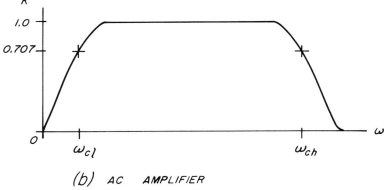

(b) AC AMPLIFIER

FIG. 7.9   Normalized frequency response of an amplifier. (a) DC; (b) AC.

phase angle versus frequency plot. These two plots define the total frequency response of the amplifier. Usually only the gain versus frequency plot is drawn. Typical plots are shown in Fig. 7.9. For convenience, the gains have been normalized to unity. Normally over the amplifier pass band ($0 \leqslant \omega \leqslant \omega_c$ for the DC amplifier; $\omega_{cl} \leqslant \omega \leqslant \omega_{ch}$ for the AC amplifier), the phase-shift angle θ is nearly constant and has the value 0° or 180°. Depending upon the number and arrangement of active elements in the amplifier, the output V_0 is either in phase or 180° out of phase with the input V_i.

If we incorporate the input network, $G_i(s)$, into the gain expression, then we can develop a relation for the ratio of the output voltage V_0 to the signal source V_s.

$$V_0/V_s = g_m G_i(s) G_0(s) = \sqrt{A_s^2 + B_s^2} \; \underline{/\theta_s}$$

The overall system gain K_s is thus

$$K_s = |g_m G_i(s) G_0(s)|$$

K_s incorporates the effect of the input network as well as the effect of the amplifier and associated output network.

7.2.2. Bandwidth

A somewhat arbitrary definition is made for bandwidth. Amplifier cutoff is defined as that frequency (ω_c) at which the power output from the amplifier is one-half of the maximum power, or

$$|V_0/V_i|^2 = 0.5|V_0/V_i|^2_{max}$$

or

$$|V_0/V_i| = 0.707|V_0/V_i|_{max}$$

since $\sqrt{0.5} = 0.707$. Amplifier bandwidth is then described by the frequency range between the 0.707 or half-power points. A DC amplifier has only an upper half-power point, whereas an AC amplifier has two as shown in Fig. 7.9. For the DC amplifier, the bandwidth (BW) is ω_c; for the AC amplifier $BW = \omega_{ch} - \omega_{cl}$.

Frequently the cutoff or half-power points are described in terms of decibel (dB) units rather than voltage units. The conversion is

$$dB = 20 \log \ (\text{voltage ratio})$$

or

$$20 \log (0.707) = -3 \text{ dB}$$

Thus the half-power points are called 3 dB points, that is signal level is 3 dB down from the maximum level. The bandwidth defined by the half-power points is then the 3 dB BW.

7.2.3. Gain-Bandwidth Product

If several amplifiers are connected in cascade, as shown for two stages in Fig. 7.10, midband gain (amplification) is increased at the sacrifice of overall bandwidth. Note that the output circuit of stage 1 (g_{m_1}) has been incorporated with the input circuit of stage 2 (g_{m_2}). The overall system gain is given by:

$$K_s = |V_0/V_s| = |g_{m_1} g_{m_2} G_i(s) G_0'(s) G_0(s)|$$

The gain-bandwidth product theorem states simply that in a single stage

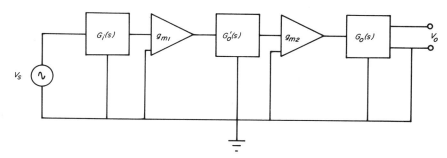

FIG. 7.10   Two cascaded amplifier stages.

amplifier, as gain is increased, bandwidth decreases such that the area under the frequency response curve (Fig. 7.9) remains constant. Simply, the product of gain and bandwidth is a constant for a given amplifier configuration. Thus for a single amplifier stage, as shown in Fig. 7.8, if we increase the value of K, BW decreases such that

$$KBW = \text{constant}$$

In multiple-stage amplifiers, as the number of stages is increased to increase gain, the bandwidth decreases to some extent generally, but not according to the gain-bandwidth–product relation.

7.2.4.   Transient Response and Risetime

Experimentally, the transient response of an amplifier can be determined by applying an input voltage V_i of the form shown in Fig. 7.11. This is called a step-voltage input and can be produced in the laboratory by a battery and switch.

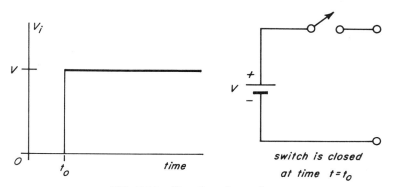

FIG. 7.11   Step-function voltage.

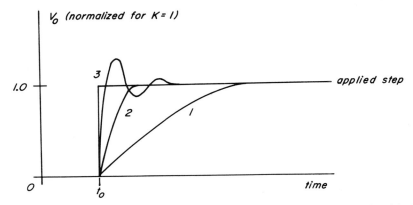

FIG. 7.12   Typical transient response showing (1) overdamped case, (2) critically damped case, and (3) damped oscillatory case.

The output voltage V_0 resulting from the application of the step voltage is displayed on an oscilloscope or chart recorder. Some typical response characteristics are shown in Fig. 7.12. Normally the response voltage V_0 will not be identical to V_i and will take one of the forms shown in the figure.

Amplifier risetime is defined in terms of response to a step-voltage input as indicated in Fig. 7.13. There are various definitions, but usually it is taken as the time required for the output voltage, V_0, to rise from 10% of the final value to 90% of the final value for an applied step input V_i. With reference to the figure,

$$\text{risetime} = t_2 - t_1 = t_r$$

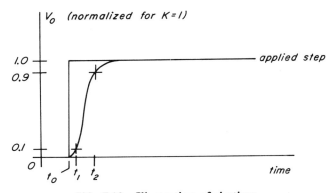

FIG. 7.13   Illustration of risetime.

Referring to the transient response presented in Fig. 7.12, we see that the underdamped condition (3) yields the fastest risetime and the overdamped condition (1), the slowest risetime.

Transient response and frequency response are directly related: the broader the bandwidth of an amplifier (at the high-frequency end), the more rapid is its transient response. Proof of this statement lies in the realm of Fourier analysis and is omitted here [see Ferris (1962), for example]. Approximately: $t_r \sim 0.35/BW$ where t_r is in seconds and BW is the 3 dB bandwidth. If we wish an amplifier to pass sharp pulses with high fidelity, then we must select an amplifier with wide bandwidth.

Any electrical signal can be represented by a summation (Fourier sum) of individual sinusoidal components of differing frequencies, amplitudes, and phase angles. It is thus possible to plot a frequency spectrum (frequency response plot, if you will) for any voltage waveform. Generally the frequency spectrum for a short pulse has the form shown in Fig. 7.14.

If the frequency response of the amplifier is not equal to the BW of the frequency spectrum of the pulse, then waveform distortion will occur.

Generally, it is impractical to use step-function test signals, and repetitive square wave input test signals are used. Figure 7.15 is a diagnostic chart for some amplifier misdesigns relative to frequency response. A general mathematical treatment of these matters will be found in Ferris (1962). A more practical treatment is given in Arguimbau and Adler (1956).

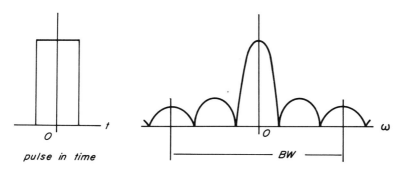

pulse in time

frequency spectrum of time pulse

FIG. 7.14   Frequency spectrum of a pulse.

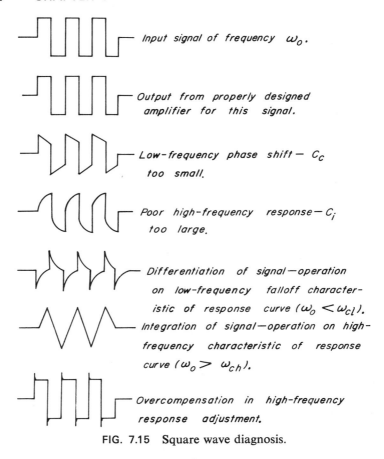

Input signal of frequency ω_o.

Output from properly designed amplifier for this signal.

Low-frequency phase shift — C_c too small.

Poor high-frequency response — C_i too large.

Differentiation of signal — operation on low-frequency falloff characteristic of response curve ($\omega_o < \omega_{cl}$).

Integration of signal — operation on high-frequency characteristic of response curve ($\omega_o > \omega_{ch}$).

Overcompensation in high-frequency response adjustment.

FIG. 7.15   Square wave diagnosis.

7.3.   AC-, DC-, AND CHOPPER PREAMPLIFIERS

The input preamplifier that is connected to a pair of electrodes used for direct physiological recording has traditionally been called a "head-stage." This terminology now appears on the wane. We have two choices of a basic amplifier system, depending upon the type of signal that is to be processed. Alternating voltages and fast pulses which possess little or no DC component are most easily amplified by an AC amplifier. Signals that are high in DC and low frequency components are processed by DC amplifiers.

There are several practical matters to be considered in selecting which amplifier to use. Theoretically a DC amplifier is suitable for all types of signals, but there are other important factors, and these are treated subsequently in this section.

It is not our purpose here to go into detailed amplifier design; the intent is simply to indicate when specific amplifier configurations are used and why. Circuits are presented only to illustrate specific points. Those interested in detailed design considerations should consult basic electronic engineering texts.

7.3.1. AC Preamplifiers

Alternating current amplifiers are recognized by the fact that input and output signals associated with the amplifier are coupled through capacitors or transformers. Capacitive coupling is generally used as normally a greater operating bandwidth is possible than with transformers. Figure 7.16 illustrates two typical input–output (coupling)

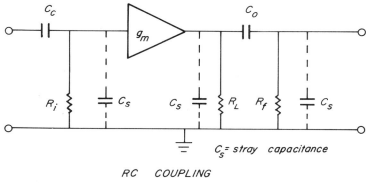

RC   COUPLING

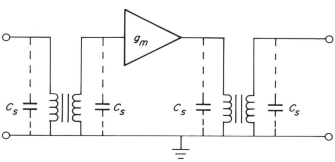

TRANSFORMER   COUPLING

FIG. 7.16   AC Amplifier coupling techniques.

techniques. In this figure, the dotted-line capacitor connections represent stray capacitance. These are important only at very high frequencies or at very high impedance levels, such as in glass micropipet recording systems. The inherent problem with AC amplifiers is in the low-frequency response. Since, normally, capacitive (RC) coupling is used, this name being evident from the figure, we will consider that case alone. Two time constants are involved: R_iC_c of the input and R_fC_0 at the output of the amplifier. The combination of R_i and C_c forms a simple high-pass filter and accounts for the low-frequency portion of the frequency response plot shown in Fig. 7.9(b). The 3 dB point at the low-frequency end is defined simply by

$$\omega_{cl} = 1/(R_iC_c) \quad \text{and} \quad f_{cl} = 1/(2\pi R_iC_c)$$

if the time constant of the output circuit is much larger than the time constant of the input circuit. The combination also acts as a differentiating circuit for the falloff portion of the filter characteristic and produces the responses shown in Fig. 7.15 for "C_c too small" and "$(\omega_0 < \omega_{cl})$." The output coupling circuit with R_f and C_0 also forms a high-pass filter and tends to increase the low-frequency falloff rate of the response curve.

The high-frequency falloff characteristic of an AC amplifier is produced by the interaction of the shunt capacitances (dotted lines) with R_i, R_f, and R_L also in RC time-constant relations, but on a parallel, rather than series, basis.

System risetime is affected by various factors. In micropipet recording systems, the RC combination of the electrode series high resistance and shunt input capacitance C_i forms a low-pass filter, which on the falloff characteristic forms an integrator circuit. Waveshape is then altered as shown in Fig. 7.15 for "poor high-frequency response" and "$(\omega_0 > \omega_{ch})$." Since there is an inverse relationship between time and frequency, poor high-frequency response corresponds to poor risetime response. The risetime of an AC amplifier is also affected by the shunt capacitance in the output circuit. The capacitance generally arises from two sources: stray capacitance to ground as a result of wiring layout, and internal capacitances associated with the active elements in the amplifier (tubes, transistors, or FETs). Risetime is equal to the product of the equivalent shunt resistance and the equivalent shunt capacitance in the output circuit, when there are no input circuit effects.

Risetime can be related to the upper 3 dB frequency cutoff point (ω_{ch}). In cases where the input circuit does not influence ω_{ch}, then risetime on a 63% basis (as opposed to the usual definition) $= 1/\omega_{ch} = R_{\parallel}C_{\parallel}$, where the values for R and C are the equivalent shunt values. If $R_{\parallel}$ is eliminated, the 63% risetime can be expressed in terms of g_m and the gain of the amplifier stage as

$$T_{63\%} = KC_{\parallel}/g_m$$

The 63% figure arises from the response characteristics shown in Fig. 7.12 for step-function response. In one time constant RC, response attains 63% of its final value.

AC amplifiers generally do not perform well at low frequencies (< 10 Hz) and exhibit signal phase shift or signal differentiation. High-frequency response must be selected so that risetime is not compromised for the signals being processed. On the other hand, AC amplifiers provide stable operation, are relatively free from baseline drift, and require much less calibration than DC amplifiers.

7.3.2.   DC Preamplifiers

Direct current (direct coupled) amplifiers are required when DC signals, slowing varying AC signals (< 10 Hz), and certain low-repetition-rate (e.g., ECG) signals are processed. In an AC amplifier, the capacitors C_c and C_o isolate the internal DC bias voltages necessary for proper amplifier operation from external DC signals. With DC amplifiers, the external DC signals are what is being processed, and since these add to or subtract from the bias voltages necessary to amplifier operation, DC amplifiers then have shifting operating points. This situation causes instability, baseline drift, and calibration difficulties. Although many instruments use DC amplifiers, their use should be avoided in some sensitive applications in favor of chopper amplifiers. Low cost is the reason that many conventional DC amplifiers are used, and many off-the-shelf instrumentation amplifiers and op-amps are very stable, thus avoiding the necessity to use chopper amplifiers.

7.3.3.   DC Offset

In order to avoid shifting the operating point of a DC amplifier beyond its linear range, it is frequently necessary to apply a DC bucking voltage at the amplifier input to offset (neutralize) the input signal. For example, suppose that we use a silver–silver chloride recording electrode and a platinum indifferent (reference) electrode in a system such as that shown in Fig. 13 of Chapter 6. In the quiescent situation at constant temperature, we can expect a small steady DC potential at the amplifier input

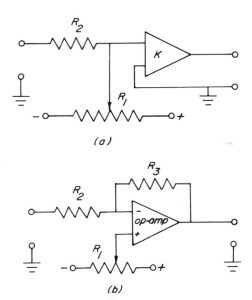

FIG. 7.17   DC offset circuits. (a) Simple technique for general amplifiers; (b) DC offset adjustment for op-amp.

terminals. A given DC amplifier might be designed to process signals which vary (+ or −) about a zero-V baseline. The electrode offset potential, in this case, would present a nonzero baseline. To correct this offset to zero, a simple circuit as shown in Fig. 7.17(a) might be used. Figure 7.17(b) shows a technique for op-amps. In commercial instruments, a more complicated system than that shown would probably be used. The "DC offset" adjustment on many commercial DC amplifiers and recorders simply provides baseline restoration for offset potentials produced by electrode systems and other sources. Use of such an offset control to restore an operating baseline does not affect amplification of signal variation about the baseline, but does introduce a varying attenuation of the input signal depending upon the value of R_1. The R_1–R_2 combination forms a voltage divider in Fig. 7.17(a).

Another problem associated with DC amplifiers (and AC amplifiers as well) is signal saturation. All amplifiers possess a certain dynamic range. This may or may not be symmetric. AC amplifiers generally have a symmetric dynamic range. DC amplifiers may be designed to accept a symmetric signal ($\pm V$ volts), or they may accept only positive signals (0 to $+V$ volts) or only negative signals (0 to $-V$ volts). Some DC amplifiers may be asymmetric in the form that they accept signals of the form $-V_1 \leqslant V_{in} \leqslant + V_2$, $|V_1| \neq |V_2|$. When the input voltage limits are exceeded, saturation generally occurs. That is, the output voltage no longer increases linearly as input is increased. In hard saturation, the output remains constant as input is increased.

Generally AC signals are symmetric about a zero-volt baseline. If a DC amplifier is used to process such a signal, it may be necessary to use a DC offset to put the AC signal into the operating range of the DC amplifier. This is also true of an AC signal that possesses a DC component.

7.3.4. Chopper Preamplifiers

Chopper preamplifiers fall into two categories: chopper or carrier preamplifiers, and chopper–stabilized preamplifiers. These devices represent a compromise between the stability of AC amplifiers and the low-pass properties of DC amplifiers. The input DC or low-frequency signal is fed through a chopper, which is simply an electrically actuated switch or optical chopper that samples the input signal at a predetermined fixed rate. The input signal is converted into a train of pulses whose amplitudes are proportional to the voltage level of the original signal. The output of the chopper is then an alternating voltage that can be amplified by a conventional AC amplifier. The AC amplifier usually is designed to have a narrow bandwidth centered about the frequency of operation of the chopper. In this manner, wide-band noise is reduced when the signal is amplified. The chopping action does inject some switching (commutator) noise into the signal.

Both mechanical and electronic choppers are used. The mechanical units frequently switch at the power line frequency (60 Hz) and through the arrangement of the switch contacts produce a chopped signal at the second harmonic (120 Hz). Mechanical choppers consist usually of a set of fixed switch contacts, an armature (moving reed with contacts), and an electromagnet assembly which actuates the armature. Although convenient, 60 Hz is not a desirable operating frequency because of possible hum pickup in the signal. Higher-frequency mechanical choppers can be designed to operate from electronic oscillators. There is, however, an upper limit on the rate at which mechanical switching can take place reliably without "contact bounce" and other problems which cause signal distortion or loss.

Electronic choppers incorporate semiconductor components in circuit configurations that are analogs of mechanical choppers. Chopper frequencies are usually on the order of 1–10 kHz, and narrow bandwidth (tuned) AC amplifiers are used. Photoelectronic choppers are

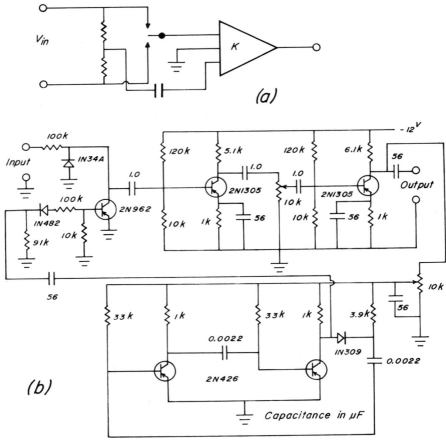

FIG. 7.18 (a) mechanical chopper-input amplifier circuit; (b) electronic chopper, multivibrator driver, and amplifier.

also used. Figure 7.18 illustrates two types of chopper input circuits; additional circuits appear in Fig. 7.19.

Chopper amplifiers do have several disadvantages. Mechanical choppers have a short lifetime and are expensive. In general, this system is satisfactory for DC signals and slowly varying AC signals. Unless quite sophisticated circuitry is used, chopper amplifiers are not satisfactory for processing repetitive pulsed signals such as ECG signals. Unless the chopping rate is very high, portions of the leading and trailing edges of such pulses, and frequently the peak value, will be lost.

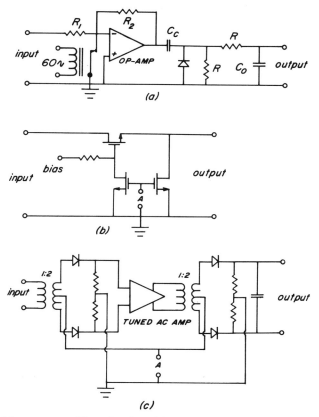

FIG. 7.19   Chopper amplifier circuits: (a) chopper amplifier using magnetically vibrating reed; (b) MOSFET chopper. The chopping signal is derived from a pulse generator or multivibrator circuit connected across "A"; (c) synchronous chopper with synchronous detector output. The driving signal is applied across "A." This circuit is also called a lock-in amplifier as the input and output circuits are driven synchronously.

After amplification, the chopped signal can be converted back to DC by using a demodulator circuit, which is simply a rectifier, and low-pass filter. Mechanical demodulators are also used and consist of additional switch contacts actuated by the same armature that chops the input signal. In this manner, synchronous operation is assured.

Advantages of chopper amplifiers, when their use is applicable, are stable operation (no baseline drift) and low noise output. Most active electronic components (transistors or tubes) exhibit $1/f$ noise, that is, electronic noise increases as frequency decreases and hence is largest at very low frequencies. If we convert a low-frequency or DC signal to the chopper frequency for amplification, then electronic noise produced in the active elements in the amplifier is greatly reduced. Operation of the chopper at 1 kHz would mean a thousand-fold reduction in noise of this origin at 1 Hz, or referred to 1 Hz.

The term carrier amplifier is sometimes applied to chopper amplifiers. The chopped waveform is the carrier signal, which has a fixed frequency; the signal being amplified then modulates the carrier. Modulation may also be achieved by frequency mixing or heterodyning rather than by chopping. This latter technique is generally used in radio-frequency applications.

Chopper-stabilized amplifiers are DC amplifiers in which a feedback signal from the output is chopped and applied at the input to stabilize the gain-drift characteristics of the DC amplifier. Typically DC amplifiers suffer from drift and other instabilities. A chopper-stabilized DC amplifier is shown in Fig. 7.20.

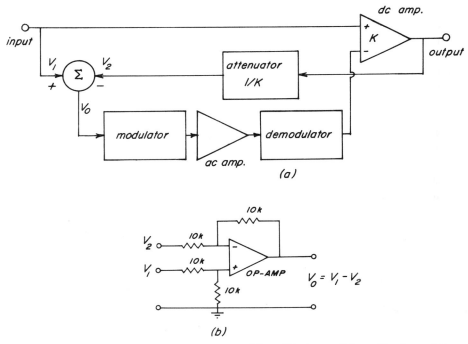

FIG. 7.20 (a) Chopper-stabilized DC amplifier. (b) A possible realization of the summing network using the op-amp circuit.

7.4. ACTIVE COMPONENTS IN PREAMPLIFIERS

The basic active devices used in preamplifiers in chronological sequence are tubes, transistors, field-effect transistors (FETs), and chips. Chips are self-contained microelectronic semiconductor circuits which are now available in numerous configurations. Thus one can buy certain basic amplifiers and other circuits as a chip for incorporation into electronic systems. Much instrument design is now being done using these ready-to-use modular units. Many chips are comparable in size to transistors and have low power requirements.

Tubes and FETs have in common the property of high input impedance. Signal input power drain is low and output coupling is enhanced by relatively low output impedance, which reduces signal loading of the amplifier stage. Tubes and FETs are generally used now for electrometer service where very high input impedances are required (see section 7.7). FETs are used in the input circuitry of differential amplifiers (section 7.5) and operational amplifiers (section 7.6).

Transistors work backwards from an impedance viewpoint. They generally present a low input impedance and high output impedance, which causes signal loading and problems with signal-output coupling. Although fine for digital work, they are being replaced by FETs and chips for much analog (continuous signal) work. Figure 7.21 is an abstract representation of a three-terminal active device and applies equally to the signal connections for a tube, transistor, or FET.

There are three basic circuit configurations in which a three-terminal active device can be used. Terminal 1 is the natural input terminal (grid of a vacuum tube, base of a transistor, and gate of an FET). Terminal 2 is the normal reference (cathode of a tube, emitter of a transistor, and source of an FET). Terminal 3 is the natural output (anode or plate of a tube, collector of a transistor, and drain of an

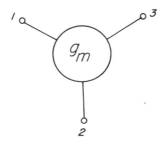

FIG. 7.21   Three-terminal active device.

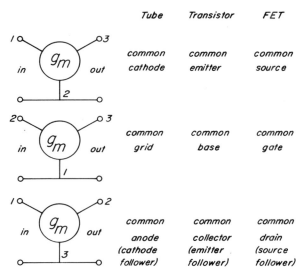

FIG. 7.22   Active element signal connections.

FET). The configuration which produces maximum voltage amplification applies the input between 1 and 2, and the output is taken between 3 and 2. In this case, the signal is shifted in phase by 180°, or "inverted." If the input is applied between 2 and 1 and the output taken between 3 and 1, some signal amplification occurs but no phase shift. When the signal input is between 1 and 3 and the output between 2 and 3, no phase shift occurs and voltage amplification is less than unity (usually about 0.9). This configuration is used for impedance conversion: high input impedance to low output impedance. Figure 7.22 summarizes these connections and indicates terminology. Methods for determining impedance levels and amplification factors, etc., using a generalized approach will be found in Ferris (1965A).

Bias techniques and design considerations are purposely omitted here. Those interested in such matters should consult standard texts on electronics, for example Brophy (1977). Detailed discussion of device operation will be found in Nanavati (1975).

7.5.   DIFFERENTIAL AMPLIFIER

The differential amplifier is a combination circuit of two simple preamplifiers into a device which either adds or subtracts two input signals. It may be represented schematically as in Fig. 7.23. Depending upon the

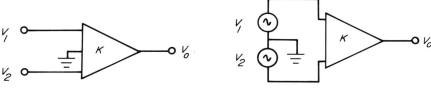

FIG. 7.23   Differential amplifiers.

internal connections in the unit, the output voltage V_0 is given by:

$$V_0 = K(V_1 + V_2) \qquad \text{summing connection}$$
$$V_0 = K(V_1 - V_2) \qquad \text{difference (usual) connection}$$

or

$$V_0 = K(V_2 - V_1)$$

The input voltages V_1 and V_2 are referenced to circuit ground. In the difference mode, this amplifier can be used to sense potential changes at one point in a circuit relative to another point in a circuit. A typical application is in ECG preamplifiers where two limb leads are connected to the input of a difference amplifier. They may also be used in null-indicator circuits to indicate equality of two voltages. When $V_1 = V_2$, then $V_0 = 0$. In this case, the voltages must be equal in both magnitude and phase angle.

7.5.1. Common Mode Signal

Let us suppose that $V_1 \neq V_2$ and that we add a voltage V_n to both. Thus, we have the augmented values $V_1' = V_1 + V_n$ and $V_2' = V_2 + V_n$. If we use V_1' and V_2' as inputs to a difference amplifier, then the output will be:

$$\begin{aligned} V_0 &= K(V_1' - V_2') \\ &= K(V_1 + V_n - V_2 - V_n) \\ &= K(V_1 - V_2) \end{aligned}$$

The voltage V_n is said to be a common mode voltage (signal). It is common to both input voltages and is subtracted out by the difference amplifier. The degree to which a difference amplifier is insensitive to a common mode voltage is called the common-mode rejection ratio (CMRR). If V_n is increased by a factor of 10,000 before a factor of one change in V_0 occurs, then the CMRR is 10,000:1.

One advantage of using difference amplifiers is that common mode signal hum (60 Hz) and noise are eliminated by the difference-input connection as long as they do not exceed the CMRR of the amplifier.

7.6. OPERATIONAL AMPLIFIERS

Originally, operational amplifiers (op-amps) were used in computers to perform mathematical operations. With the development of semiconductor integrated circuits, small and inexpensive op-amps are now available and are being used in place of conventional discrete component amplifiers in many applications. Basically an op-amp is a very high gain (theoretically infinite gain) amplifier. In use, it is normally connected in a feedback mode of operation. That is, a portion of the output voltage is returned to the input terminals with 180° of phase shift such that the returned voltage subtracts from the input voltage. In this manner, amplifier gain is both controlled and stabilized. The basic network configuration is shown in Fig. 7.24. Figure 7.25 shows a detailed circuit for an op-amp.

The advantage of op-amps is that they can be used as building blocks to construct a number of useful circuits. In most cases, only resistors are needed to complete a design. We will now examine some particular circuits. At this point, the actual arrangement of active devices to make an op-amp is of no concern to us. We will simply use the device as it comes from the manufacturer.

The op-amp connection to make a follower circuit (emitter follower, etc.) is shown in Fig. 7.26.

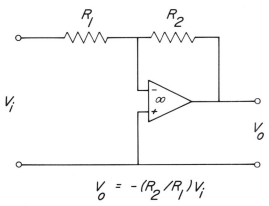

$$V_o = -(R_2/R_1)V_i$$

FIG. 7.24   Basic operational amplifier circuit.

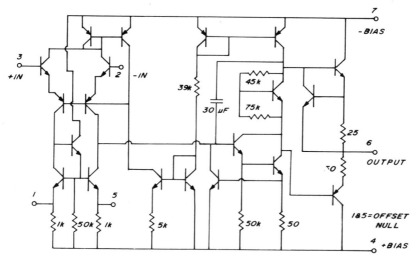

FIG. 7.25 Circuit schematic for operational amplifier chip (National Semiconductor type LM741/LM741C).

Another useful circuit is the noninverting amplifier that provides gain without phase shift. Circuit connections using an op-amp are shown in Fig. 7.27.

If 180° phase shift is desired (inverting amplifier), then the signal connections are as shown in Fig. 7.28.

A single op-amp may be used as a differential amplifier by connecting it as shown in Fig. 7.29.

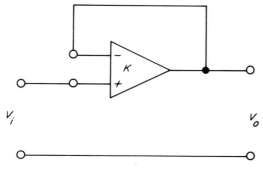

FIG. 7.26 Operational amplifier voltage follower. The mathematical relations that apply are: $V_0 = (V_i - V_0)K$, which may be rewritten as $V_0 = V_i - (V_0/K)$, a form which more clearly indicates that $V_0 = V_i$ as $K \rightarrow \infty$.

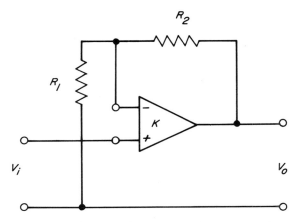

FIG. 7.27   Noninverting operational amplifier: $V_0 = V_i(R_1 + R_2)/R_1$.

In this section, we have discussed signal connections only for operational amplifiers. It is also necessary to provide DC bias voltages. A single-polarity DC supply may be used with two options: 1. a positive voltage is connected to the positive bias input, and the negative bias input is grounded; 2. a negative voltage is connected to the negative bias input, and the positive bias input is grounded. Normally a bipolar supply is used with the bias terminals connected to plus and minus supplies of the same magnitude. An op-amp may also be used to pro-

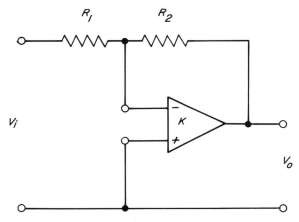

FIG. 7.28   Inverting operational amplifier: $V_0 = -(R_2/R_1)V_i$. To improve stability, the + terminal is frequently connected to ground through a resistor $R_3 = R_1R_2/(R_1 + R_2)$, rather than being directly connected as shown in the diagram above.

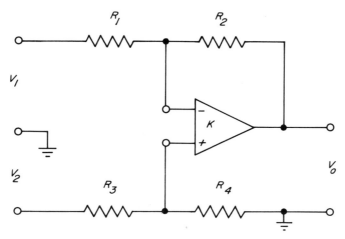

FIG. 7.29 Differential-input-mode operational amplifier: $V_0 = (R_2/R_1)(V_2 - V_1)$ when K is large and $R_1 = R_3$, $R_2 = R_4$, otherwise $V_0 = V_2(R_4/R_1)(R_1 + R_2)/(R_3 + R_4) - (R_2/R_1)V_1$.

duce a bipolar supply from a single battery, as shown in Fig. 7.30. The follower circuit generates a reference ground while drawing only 10 μA from the battery if an RCA 3078 op-amp is used. When operated symmetrically, this circuit produces plus and minus voltages equal to one-half of the battery voltage.

The output signal voltage swing must not exceed the total bias voltage or signal clipping and saturation will occur. For an op-amp biased with a typical value of ± 15 V DC, the peak-to-peak signal voltage swing is 30 V. DC offset voltages present in the signal input can drive an op-amp to saturation, although the AC signal swing may within acceptable limits.

Operating bias voltages vary and must be determined from the

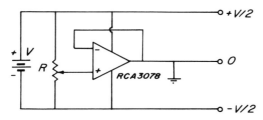

FIG. 7.30 Use of an operational amplifier to produce a bipolar power supply with common ground.

specification sheets for a particular device. Voltages of the order of 15–20 V DC are typical.

The connections which we have shown are for DC input. If AC operation is desired, all that is necessary is the insertion of a coupling capacitor in series with the input signal lead.

We have not discussed values of resistance to be used with op-amp circuits. In most cases, the manufacturer will indicate minimum and maximum values. If we examine the voltage gain expression for the inverting amplifier

$$V_0 = -(R_2/R_1)V_i$$

we see that we can adjust gain simply by selecting R_1 and R_2 in the correct ratio. Thus if we require an amplifier with a gain of 100, then $R_2/R_1 = 100$ or $R_2 = 100\ R_1$; for $R_1 = 1\ k\Omega$, $R_2 = 100\ k\Omega$.

Stray capacitance influences the operation of op-amps with respect to time constants and overall response. When large resistance values are used, the resulting RC time constants produced by stray capacitance may degrade amplifier performance to unacceptable levels. In some cases, compensation can be provided by placing a small capacitance (usually < 50 pF) in parallel with the feedback resistor. This technique, however, may lead to oscillation in some circuits. When the resistance values are too small, undesirable loading of the op-amp may occur with resulting poor performance.

There are various factors, in addition to biasing, which must be considered when using op-amps. Op-amps will not function above a predetermined value for V_0. When V_0 exceeds the stated (specification sheet) value, saturation occurs. Since op-amps can be operated in both inverting and noninverting modes, saturation is normally symmetric and is determined by $V_0 = \pm V_{sat}$.

Open-loop gain is defined for the basic op-amp without any feedback connections (return connections, either direct or through resistors, from output to input), and typical values for open-loop gain are $> 100,000$. Closed-loop values for gain are determined from the expression relating V_0 and V_i in the figures, and depend upon resistance ratios.

Other factors are op-amp bandwidth and compensation. Generally op-amps are low-frequency devices. Unless an input DC blocking capacitor is used in the signal path, they respond to zero frequency (direct current). The high-frequency 3 dB point is usually in the range 1–10 kHz. Bandwidth depends upon how much gain is desired. The higher the gain, of course, the lower the BW. A typical op-amp characteristic is plotted in Fig. 7.31. From the figure, we note that for a gain of 1000

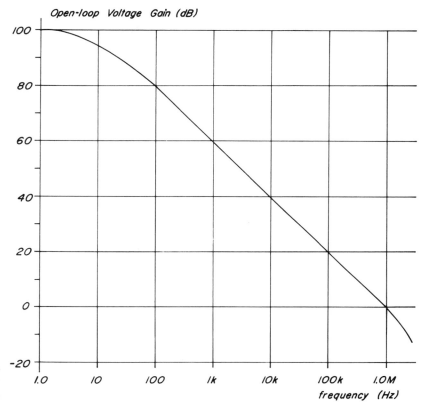

FIG. 7.31   Gain-bandwidth characteristic for a typical operational amplifier: voltage gain (dB) $= 20 \log(V_{out}/V_{in})$.

(60 dB), the bandwith is 1 kHz. If the gain is reduced to 10 (20 dB), the bandwidth is 100 kHz.

Op-amps generally contain compensation networks to increase stability and external bandwidth. In some cases, these are adjustable. Again, the manufacturer's specification sheets provide this information. Other factors to be considered in using op-amps are noise, drift in operating point, and temperature stability. These vary with different op-amp configurations. Good general references for these problems are handbooks supplied by manufacturers of op-amps (Burr-Brown 1963, for example, and Analog Devices, Fairchild, Precision Monolithics, etc.).

Op-amps function quite well in performing the operations of integration and differentiation. Basic circuits are shown in Fig. 7.32.

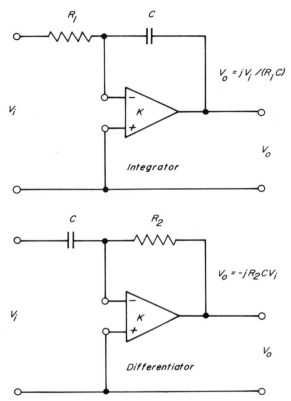

$$V_o = jV_i/(R_1C)$$

Integrator

$$V_o = -jR_2CV_i$$

Differentiator

FIG. 7.32   Operational amplifier integrator and differentiator circuits.

7.7.   ELECTROMETER PREAMPLIFIERS

Electrometer preamplifiers are used where very low current drain from a signal source is important. Input impedances of such circuits lie in the range from 10^9 to 10^{14} Ω. Typical applications are preamplifiers for microelectrode recording, and null sensing circuits in potentiometric devices such as pH meters. The active elements in these circuits are either electrometer vacuum tubes or insulated gate FETs. Ordinary transistors are inapplicable because of their low input impedance.

When vacuum devices are used, either tubes designed especially for electrometer service can be employed, or certain miniature pentode tubes can be connected for electrometer operation. Normally electrometers are used as DC signal amplifiers when very small currents are available. Because of their high-input impedance, their time constants are long and they do not generally respond rapidly to fast transients nor do they recover quickly from signal overload. Vacuum electrometers are sensitive to a number of environmental factors and require much care in use. FETs are more rugged in this respect, although one must be careful that excessive charge buildup does not occur on the gate terminal, since this can destroy an insulated gate FET.

Electrometer circuits *per se* are not directly applicable to physiological recording systems because of their long time constants. Even if the shunt capacitance of the input circuit is only 1 pF, for the range of input shunt resistance from 10^9 to 10^{14} Ω, the time constants of the input circuit will vary from 1 ms to 100 s. These time constants are excessive. Where electrometer techniques do find application is in special feedback circuits designed to cancel input capacitance. These are discussed below.

7.7.1. Negative-Input-Capacitance Preamplifier

In 1949, Bell published a circuit for use with certain video amplifier systems. The system is an adjustable self-neutralizing feedback amplifier, which, when properly adjusted, has no shunt input capacitance. A functional diagram is shown in Fig. 7.33. The input capacitance of the system shown in Fig. 7.33(a) is developed from

$$V_0 = KV_i$$

$$i = C_f \frac{d}{dt}(V_i - V_0) = C_f \frac{d}{dt}(V_i - KV_i)$$

$$= C_f \frac{d}{dt}(V_i)(1 - K)$$

where i is the current in C_f with a sense from input to output. But

$$i = C_{in}dV_i/dt$$

therefore

$$C_{in} = (1 - K)C_f$$

If $K > 1$, then the effective input capacitance C_{in} is negative and will cancel a positive (real) shunt capacitance at the input terminals. Referring to Fig. 7.33(b), we want to satisfy the condition

$$C_i + C_{in} = 0$$

or

$$C_i = -C_{in} = -(1 - K)C_f$$

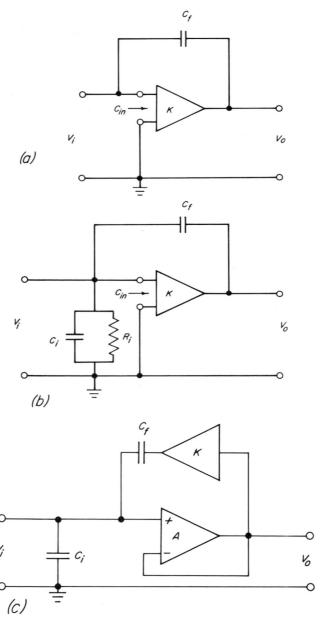

FIG. 7.33 Functional diagram for negative-input-capacitance amplifier: (a, b) diagrams for basic calculations shown in text; (c) functional diagram for the realization of this technique (kindly suggested by Dr. T. R. Colburn, NIH, Bethesda, Md.)

184

or

$$C_i = (K - 1)C_f$$

An appropriate value of K will permit cancellation of C_i, but with the constraint from above that $K > +1$.

Figure 7.33(c) presents a practical realization technique. Amplifier A provides a fixed-gain output for measuring V_i. In the diagram, it is shown in the follower mode with unity gain, but it may have any

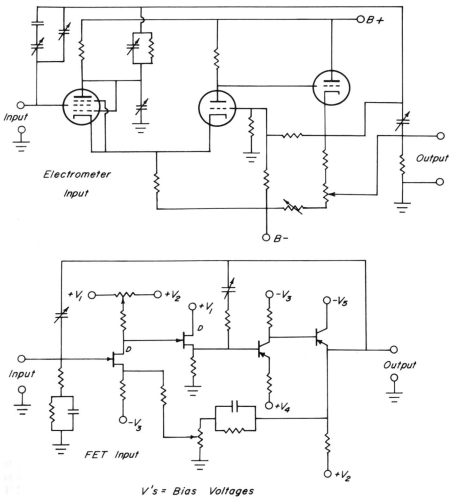

V's = Bias Voltages

FIG. 7.34   Basic circuits for negative-input-capacitance amplifiers.

positive gain value. A second amplifier with variable gain K provides capacitive feedback to the positive input terminal of amplifier A. When $(K - 1)C_f = C_i$, the system input capacitance is neutralized, and C_f should be chosen approximately equal to C_i, so that K can be varied over the range $+1 < K < +3$.

Additional discussion of neutralization techniques has been presented in some detail in Millman and Taub (1956). Their analysis relates in particular to operational amplifier and integrator circuits.

Several practical designs have evolved from the functional representation shown in Fig. 7.33. These have been described by Amatniek (1958) and Bak (1958). Generalized schematic diagrams are presented in Fig. 7.34. Electrometer-tube input circuits are preferable to transistor input, although field effect transistors are almost as good. Input resistances of the order of 10^{12} Ω are possible with tubes and of the order of 10^{11} Ω with FETs in the feedback configurations indicated.

A modified form of the Bak unit has been used by the author, and performance curves are shown in Fig. 7.35. The effect of the setting of the neutralizing capacitor upon transient response and signal waveshape is shown.

In the design of negative-input-capacity amplifiers, one must insure that tube grid current or gate signal current is minimized. These currents can have a deleterious effect upon fidelity of measurement: by modifying the properties of the physiological system (such as a single cell), causing baseline drift, polarizing the electrodes, causing a nonlinear variation in amplifier input resistance and concomitant change in

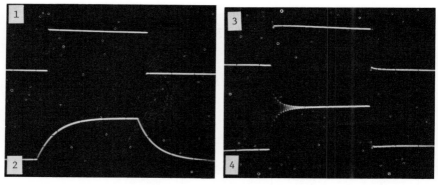

FIG. 7.35 Transient response of negative-input-capacitance amplifier. Traces: 1. undistorted source signal; 2. distorted signal as a result of cable shunt capacitance (measured at amplifier input but without compensation); 3. output signal from amplifier after compensation adjustment; 4. output signal from amplifier with "ringing" because of overcompensation adjustment.

signal amplitude, generating noise. A safe grid current with regard to these considerations is one that is less than 10^{-12} A. The use of fluid-bridge electrodes lowers the probability that the electrodes will be polarized. A higher probability of electrode polarization exists where metal electrodes come into direct contact with the physiological system.

7.7.2. General Properties of Some Electrometer Devices

Table 7.1 indicates and intercompares properties of some electrometer devices. Other data are to be found in Aronson (1977). Additional properties of concern are: overload capacity before the device is damaged, restabilization time after overload, and microphonics. FETs generally exhibit very low noise arising from mechanical shock or vibration, since, when correctly packaged, there is little to move or vibrate. These devices are easily damaged by overload currents, so that restabilization time is not a factor. Because of the way in which they are fabricated (metal electrodes supported in a vacuum), electrometer tubes tend to produce substantial microphonic noise. Overload

Table 7.1

Electrometer Properties

Property	MOSFET	JFET	Electrometer vacuum tube
Input resistance	$\sim 10^{14}\,\Omega$	$\sim 10^{12}\,\Omega$	$\sim 10^{14}\,\Omega$
Input capacitance (minimum)	0.5 pF	2 pF	5 pF
rms Noise			
1. voltage			
0.1–10 Hz	10 μV	5 μV	5 μV
10–500 Hz	100 μV	12 μV	30 μV
2. current			
0.1–10 Hz	5 fA	20 fA	5 fA
Stability			
1. voltage			
time (/24 h)	1 mV	50 μV	4 mV
temp (/°C)	150 μV	10 μV	0.5 mV
2. current			
time (/24 h)	$\sim$1 fA	$\sim$1 pA	$\sim$1 fA
temp (/°C)	$\sim$1 fA	(X2/10°C)	$\sim$1 fA
Input offset current	$\sim$5 fA	$\sim$1 pA	$\sim$20 fA

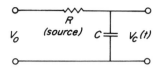

FIG. 7.36   Electrometer input circuit.

behavior, however, is good and restabilization time depends upon the
RC time constant of the input circuit (essentially the time required for
the overload charge to leak off the grid). Popular electrometer tubes are
the British Z729 pentode and the American CK5702 pentode.

As shown in Fig. 7.36, the input circuit of an electrometer device
has a simple RC time constant. If a step function of voltage is applied,
the voltage across the capacitor is given by

$$v_c(t) = V_0(1 - e^{-t/RC})$$

The capacitor charging characteristic as a function of time constant
($\tau = RC$) is given in Table 7.2. From the table we can see that if we
have a device in which the source resistance is very high (such as a glass
potentiometric electrode) and associated cable capacitance associated
with the interconnecting cable between the source and the electrometer,
the response time can be quite long. Typical shunt capacitance for
shielded cable is 22 pF/ft. If we consider a 10^{12} Ω source driving a foot
of shielded cable connected to an electrometer device with 2 pF input
capacitance, the associated time constant is

$$\tau = (10^{12})(24 \times 10^{-12}) = 24\text{ s}$$

The time required to reach 99% of final value is

$$t = (4.605)(24) = 110.52\text{ s} = 1.84\text{ min}$$

Table 7.2

Electrometer Capacitor Charging Characteristic

Time expressed in multiples of τ	Capacitor charge as % of voltage applied
0	0
1	63.2
2	86.5
3	95.0
4	98.2
4.605	99.0
5	99.3

Table 7.3

Electrometer Overload-Charge Leak Characterstic

Time expressed in multiples of τ	Device overload charge as % of initial overload charge
0	100
1	36.8
2	13.5
3	4.98
4	1.83
4.605	1.00
5	0.67

By similar argument, the time required for an overload charge to leak away from an electrometer device is determined by the expression

$$q(t) = Q_0 e^{-t/RC}$$

Thus from Table 7.3 we can see that for a device in which the input resistance and capacitance are, respectively, $10^{14}\Omega$ and 5 pF, the time constant is 500 s. It would then require 2300 s or > 38 min for an overload charge to leak away from the device. Normally some sort of overload reset device is required. Either a momentary switch that shorts the input to ground or a switched low-value bleeder resistance may be used. The long time constants associated with electrometers generally limit bandwidth to a few Hertz.

Another electrometer device is a vibrating capacitor, as shown in Fig. 7.37. One plate of the capacitor is fixed, and the other is vibrated by an AC coil or solenoid. Generally the device is called a vibrating-reed capacitor and operation is similar to some mechanical choppers, as mentioned in section 7.3. The charge on the capacitor is converted into

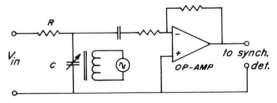

FIG. 7.37 Circuit for a vibrating capacitor electrometer: R is an isolating resistor placed between the signal source and the vibrating capacitor C; the vibrating plate of the capacitor is driven magnetically by a coil connected to a sinusoidal oscillator.

a proportional AC voltage signal. For mechanical reasons, the frequency of vibration of the reed is limited to 100 Hz or less, which limits device bandwidth to a few Hertz. These devices, however, exhibit an input impedance greater than 10^{16} Ω. Minimum input capacitance is 2 pF, rms noise voltage is approximately 1 μV, current stability is 500 times better than in electrometer tubes, and voltage stability is similar to a JFET. Typical applications are in amplifiers for ionization chambers, ion-specific electrodes (high impedance), and various industrial circuits.

Electrometer devices, in general, find applications in circuits associated with certain photodetectors, ion-specific and pH electrodes, biomedical recording electrodes, low-current measuring devices, and high-source-impedance transducers. They serve in applications where a very high input impedance is required and long time constants with narrow bandwidth are unimportant.

7.8. LOGARITHMIC AMPLIFIERS

As will be seen from the subsequent discussion of the Beer–Lambert law in Chapter 10, outputs from transducer systems may vary exponentially or logarithmically with excitation, rather than linearly. This is especially true in certain types of optical systems. Although indicating instruments can be calibrated with nonlinear scales (such as in a conventional ohmmeter), it is generally desirable to have linear displays to insure uniformity in reading accuracy of the scales. Linearization of a logarithmic or exponential response can be achieved through the use of logarithmic amplifiers (log-amps).

The basic log-amp circuit is shown in Fig. 7.38 in which a matched silicon transistor pair is used (2N2453, 2N2920). The op-amps may be 741s or other general purpose op-amps. The output voltage V_0 as a function of input voltage V_i is given by

$$V_0 = -\frac{R_2 + R_3 + R_6}{R_2 + R_3} \gamma_T \log \frac{V_i}{R_1 i_r}$$
$$= -K\gamma_T \log A V_i$$

where i_r is the reference current, γ_T is the temperature-scaling factor, K is the gain of follower op-amp, and A is the dynamic range adjustment to obtain a useful dynamic range $\sim 10^7$. R_2 is a temperature-dependent resistor used to stabilize amplifier gain, and

$$\gamma_T = 2.3\eta V_T$$

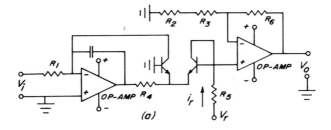

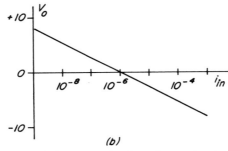

FIG. 7.38   Basic log-amp circuit and transfer characteristic.

where V_T is the temperature equivalent of voltage

$$V_T = kT/e = T/11{,}600$$

where k is the Boltzmann constant, e is the magnitude of electron charge, T is the Kelvin temperature, and η is an empirical parameter ~ 1 for germanium and ~ 2 for silicon. Thus for a silicon device operating at 300 Kelvin

$$\gamma_T = (2.3)(2)(300/11{,}600) = 0.12 \text{ V}$$

The circuit shown in Fig. 7.38 may be used to linearize a response function which changes exponentially with excitation, subject to scale factors (K and A) of the circuit.

Another circuit which accomplishes the log function response is shown in Fig. 7.39. This is actually a log-ratio amplifier, but if the input V' is set to 1.0 V, then

$$V_0 = \frac{R_3}{R_2} \gamma_T \log V_i$$

otherwise,

$$V_0 = \frac{R_3}{R_2} \gamma_T \log (V_i/V')$$

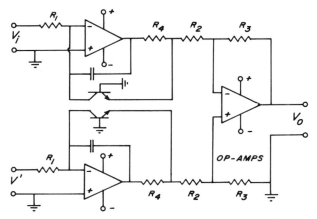

FIG. 7.39   Log-ratio amplifier circuit; this acts as a log-amp when $V' = 1V$ DC.

To linearize a logarithmic excitation–response relation, the antilog circuit of Fig. 7.40 can be used. The voltage transfer relation is

$$V_0 = R_3 i_r \log^{-1}\left[\frac{-R_2 V_i}{(R_1 + R_2)\gamma_T}\right]$$
$$= K' \log^{-1}(-A'V_i/\gamma_T)$$

where $K' = R_3 i_r$ and $A' = R_2/(R_1 + R_2)$.

Teledyne Philbrick manufactures two logarithmic modules that can be used for implementing these circuits. Model 4357 responds to positive input signals and model 4358 responds to negative inputs. The dynamic range of each is six decades for current and four decades

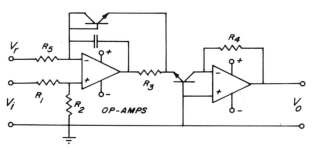

FIG. 7.40   Antilog-amp circuit.

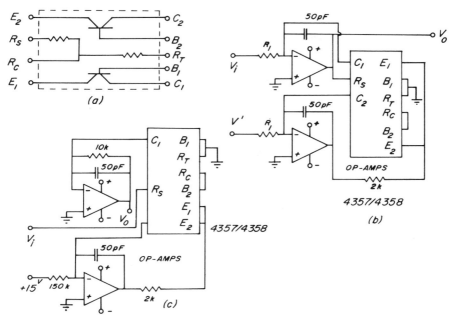

FIG. 7.41   Teledyne Philbrick log-amp circuits: (a) basic TP 4357/4358 module; transistor polarity depends upon model number; (b) module used in log-ratio amplifier circuit; (c) module used in antilog-amp circuit.

for voltage. The basic package, log-amp, and antilog-amp circuits are shown in Fig. 7.41.

An application of log–amps and other techniques is presented in the circuits shown in Fig. 7.42, and is the schematic for the analog portion of a scanning densitometer. This device is used to scan the optical density of colored bars or dots produced on electrophoresis plates and thin layer or paper chromatograms. The system uses a pulsed light source with a 1 kHz carrier frequency. The light which passes through the plates is sensed by a United Detector Technology sensor, amplified, and passed through a 1 kHz band-pass filter/amplifier (Budak, 1974). The filter output passes through an op-amp follower used for impedance matching and the optical signal is recovered using a precision rectifier (detector) circuit. Linearization of the optical excitation–response (Beer's Law) is accomplished by the log-amp circuit, the output of which drives a paper chart recorder.

An electrophoresis plate (human blood serum) and the strip chart output from the densitometer are shown in Fig. 7.43(a) and (b), respectively. Albumin and the globulin fractions appear in the tracing.

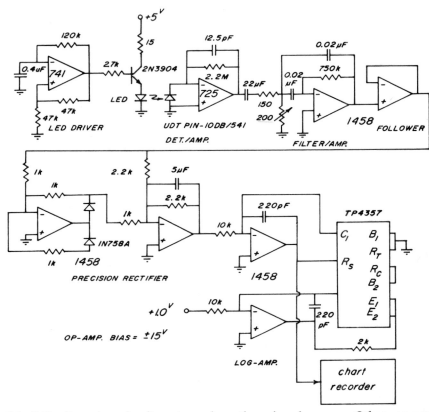

FIG. 7.42 Scanning densitometer schematic using log-amp. Other op-amp circuits and the United Detector Technology silicon diode discussed in the text are also incorporated into the design. If a DC offset adjustment (baseline adjustment) is required, the amplifier circuit of Fig. 7.17(b) may be connected following the log-amp circuit. The "chart recorder" connection would be connected to either the + or − terminal of the op-amp to obtain noninversion or inversion of signal polarity. Scanner drive motor circuitry and other controls are not shown.

7.9.   FEEDBACK

The circuits described in sections 7.6–7.8 have depended upon feedback. Amplifiers have been described in which part of the output signal was returned to the input. Feedback is used to stabilize amplifier operation or to give a controlled output. Figure 7.44 illustrates a basic feedback amplifier. The output–input relation is:

$$V_0 = KV_i + KG_f(s)V_0 = \frac{K}{1 - KG_f(s)}\, V_i$$

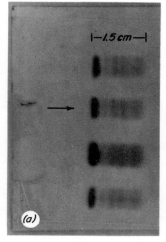

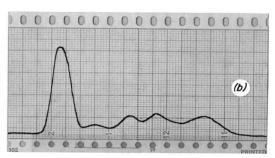

FIG. 7.43   Performance of scanning densitometer: (a) human blood serum electrophoresis plate (Gelman Sepratek™ method); (b) trace from scanning densitometer. In (a), the arrow indicates the record scanned.

$G_f(s)$ may be a simple resistor network or it may be a frequency selective network. If $G_f(s)$ is simply a resistor network, the effect of the feedback will be to reduce the overall gain of the amplifier, since

$$\text{amplifier gain} = \frac{V_0}{V_i} = \frac{K}{1 - KG_f(s)}$$

For a resistive network, $G_f(s)$ is a number which may be > or < unity.

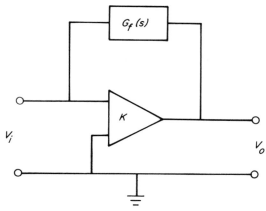

FIG. 7.44   Basic feedback amplifier.

The feedback in this case produces gain stabilization. Amplifier frequency response is increased (gain–bandwidth relation) and stability is improved. There is also a reduction in electrical noise at the output because of signal cancellation.

When $G_f(s)$ is frequency selective, then phase shift occurs in the signal that passes through the feedback network. If the return signal is out of phase with V_i, then signal subtraction occurs and the feedback is said to be negative. If the return signal is in phase with V_i, then positive feedback occurs. Negative feedback leads to stabilized operation and positive feedback to unstable operation. Positive feedback through a very frequency-selective network provides the basis for tuned (frequency selective) amplifiers and oscillators. Positive feedback increases selectivity (decrease in BW), increases gain, and promotes instability. Negative feedback decreases selectivity (increase in BW), decreases gain, and promotes stability.

Feedback is also used to provide a constant (controlled) voltage as shown in Fig. 7.45, in which V_0 is the controlled voltage. The output is sampled by the voltage divider network of R_1 and R_2, and the feedback voltage V_f is compared against a reference V_i.

$$V_0 = K(V_i - V_f)$$

$$V_f = \frac{R_2}{R_1 + R_2} V_0$$

$$V_0 = \frac{K(R_1 + R_2)}{R_1 + R_2(1 + K)} V_i$$

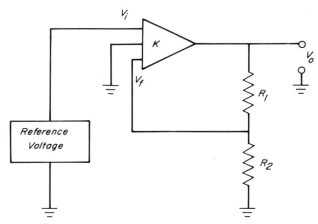

FIG. 7.45   Controlled-output feedback system.

Initially V_i and the ratio $R_2/(R_1 + R_2)$ are adjusted to give a particular V_0. If V_0 then starts to change, V_f changes such that the amplifier input $(V_i - V_f)$ changes, which then returns V_0 to its preset value. The error signal applied to the amplifier is:

$$E = V_i - V_f$$

$$= \frac{R_1 + R_2(1 + K)}{K(R_1 + R_2)} V_0 - \frac{R_2}{R_1 + R_2} V_0$$

$$= V_0/K$$

Thus the error E is minimized if K is large. For example, let $V_0 = 100\text{ mV}$ (typical value in cellular recording) and let the permissible error be 0.1, then $K = 1000$.

A full treatment of feedback systems is beyond the scope of this text, and the reader is referred to the literature [see Millsum (1966) for example].

7.10. ISOLATION NETWORKS

In excitation–response systems such as are used in neurophysiological studies and some impedance plethysmography techniques, a stimulus artifact can result from the stimulating current reaching the recording electrodes directly. In some instances, this can be desirable when propagation time measurements are being made. In most physiological work, stimulus artifact is undesirable. Artifact signals are generated when both the stimulating and recording systems have a common ground. To avoid this situation, the stimulator is isolated from the ground system of the recording network by an isolation network that generally contains a pulse transformer as shown in Fig. 7.46.

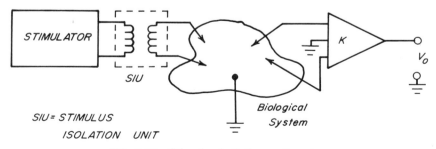

FIG. 7.46   Stimulus isolation unit system.

In this system, a differential amplifier is used, and its reference ground is the same as the ground reference in the physiological system being studied. The stimulus unit (pulse generator) is floating with respect to the rest of the system.

Generally speaking, the most stable operation of stimulus–response systems occurs when constant current signal sources are used. These can be developed through the use of op-amps as shown in Fig. 7.47. Optical isolators, as discussed in Chapters 3 and 12, are popular and in widespread use. They eliminate the need for transformers and for the oscillators used in DC/DC converters, as indicated in Fig. 7.48.

In some instances, a more sophisticated stimulus isolation unit is used, as is shown in block diagram form in Fig. 7.48 (Strong, 1970).

Typical performance values for this system are: isolation from constant current source, 5 pF coupling at $> 10^{10}$ Ω; isolation of output from ground, 30 pF coupling at $> 10^{10}$ Ω.

7.11.   NOISE

Amplifier systems generate electrical noise, which adds to the signals being processed. The source of this noise is the passive elements in the circuitry and the active elements themselves. Other noise sources are electrodes and environmental electrical noise from switchgear, motors, fluorescent lamps, etc., and vibrational or mechanical noise (microphonics).

Noise produced by active devices (tubes, transistors, FETs) is usually $1/f$ type. That is, the noise voltage produced in the device is inversely proportional to frequency. This type of noise affects DC and low-frequency AC amplifiers in particular.

Noise produced in passive circuit elements (resistors especially) is usually of thermal origin (white or Johnson noise) and has the mean-square voltage value:

$$v^2 = 4kTR \, \Delta B$$

where k is Boltzmann's constant, T is the Kelvin temperature, R is the equivalent resistance of the noise generating network, and ΔB is the bandwidth of the instrument used to measure v^2. Thermal noise is most apparent in wideband (AC) amplifier systems.

$$v_{\mathrm{rms}} \sim 1.3 \times 10^{-10} \sqrt{R \, \Delta B} \ \mathrm{V}$$

If $\Delta B = 100$ kHz and $R = 10$ kΩ, then $v_{\mathrm{rms}} \sim 4.11$ μV.

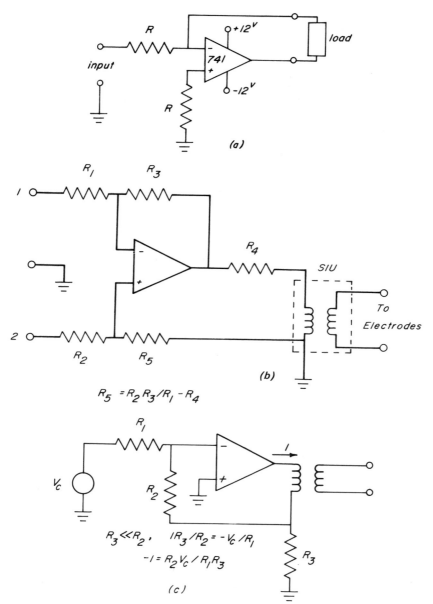

FIG. 7.47 Constant current sources: (a) simple system; R is selected to match output impedance of signal source; (b) more complex system which accepts bipolar inputs; (c) the more usual practical configuration. The current I is converted to a voltage in the small sampling resistance R_3. The op-amp maintains that voltage in correspondence with the command voltage V_c. Circuit (c) and analysis courtesy of Dr. T. R. Colburn, National Institutes of Health, Bethesda, Md.

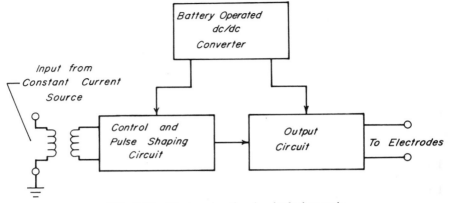

FIG. 7.48   Electronic stimulus isolation unit

The bandwidth of the measuring or responding device varies widely. For some mechanical indicating meters, $\Delta B = 1$ Hz and all noise is rejected except for the frequency component to which the meter is tuned. An RC input circuit for a device (source resistance and total input capacitance) yields a bandwidth of $1/(4RC)$. For an instrument with a risetime t_r (10–90%), $\Delta B = 0.55/t_r$.

An overall noise figure (NF) can be defined for an amplifier or a system. Several definitions exist including one in terms of signal-to-noise ratio (S/N):

$$NF = \frac{\text{input } S/N}{\text{output } S/N}$$

$$= \frac{\text{real output noise}}{\text{output noise from noisefree amplifier}}$$

$$= \frac{\text{total } v^2 \text{ noise output}}{\text{source } v^2 \text{ noise}}$$

Both S and N represent powers (v^2) and not voltages. Thus noise levels must be added in quadrature, for example:

Total v^2 output noise $=$ source v_s^2 noise
$+$ instrument (amplifier) v_i^2 noise

7.12.   REFERENCES

Aronson, M. H., 1977, "Low-level amplifiers: Part 6. Electrometers," *Med. Electron. Data* **8**, C1–C15.

Amatniek, E., 1958, "Measurement of Bioelectric Potentials with Microelectrodes and Neutralized Input Capacity Amplifiers," *Trans. IRE, PGME* **10**, 3.

Arguimbau, L. B., and R. B. Adler, 1956, *Vacuum-tube Circuits and Transistors*, Wiley, New York.

Bak, A. F., 1958, "A Unity Gain Amplifier," *EEG Clin. Neurophys.* **10**, 745.

Bell, P. R., 1949, "Negative Capacity Amplifier, Appendix A," pp. 767–770, in *Waveforms* (B. R. Chance *et al.*, Eds.). MIT Radiation Laboratory Series, McGraw-Hill, New York.

Brophy, J. J., 1977, *Basic Electronics for Scientists*, 3rd Ed., McGraw-Hill, New York.

Budak, A., 1974, *Passive and Active Network Analysis and Synthesis*, Houghton Mifflin Co., Boston.

Burr-Brown, 1963, *Handbook of Operational Amplifiers*, Burr-Brown Research Corp., Tucson, Arizona.

Ferris, C. D., 1962, *Linear Network Theory*, Merrill, Columbus.

Ferris, C. D., 1965A., "A General Network Representation for Three-Terminal Active Devices," *Trans. IEEE*, E–8(4), 119.

Ferris, C. D., 1965B, "Low Frequency Bridge Detector," *Rev. Sci. Instr.*, **36**(11), 1652.

Ferris, C. D., 1974, *Introduction to Bioelectrodes*, Chapter 7, Plenum, New York.

Malmstadt, H. V., C. G. Enke, and S. R. Crouch, 1974, *Electronic Measurements for Scientists*, Benjamin, Menlo Park, California.

Millman, J., and H. Taub, 1956, *Pulse and Digital Circuits*, McGraw-Hill, New York.

Millsum, J. H., 1966, *Biological Control Systems Analysis*, McGraw-Hill, New York.

Nanavati, R. P., 1975, *Semiconductor Devices: BJTS, JFETS, MOSFETS, and Integrated Circuits*, Intext, New York.

Strong, P., 1970, *Biophysical Measurements*, Tektronix Corp., Beaverton, Oregon.

8

Detectors for Transducer Systems

We have discussed various electromechanical, acoustic, and optical transducers as well as electrodes. In many instances, the transducers produced an electrical output that could be directed to an amplifier without additional signal conditioning. In other cases, the transducer output was not simply a voltage proportional to the transducer excitation. It is these devices that we examine in this chapter.

8.1. VARIABLE CAPACITANCE AND INDUCTANCE DEVICES

As noted in Chapter 3, bridge techniques are not generally suitable for variable capacitance or inductance transducers. Several factors are responsible for this, and include such matters as stray capacitance in cabling (usually the basic transducer C or L values are very small), strays in bridge circuits, unequal phase-angle changes in the arms of the bridge, and very small changes in C or L. For these reasons, it is more

202

appropriate to use the variable C or L to shift the frequency of an electronic oscillator. This technique, however, presents the problem of converting a frequency shift to a variable voltage. Two methods by which this can be achieved are the oscillating detector or autodyne circuit and the linear discriminator.

8.1.1.   Oscillating Detector or Autodyne Circuit

The basic circuit is illustrated in several forms in Fig. 8.1. In principle, it operates as follows: The circuit represents a high-gain tuned amplifier which oscillates at a very low level (marginal oscillator). The parallel LC combination in the input circuit determines the frequency of oscillation. Changes in C or L produced by excitation applied to the transducer cause a shift in the oscillator frequency. Because of the marginal amplifier–oscillator nature of the circuit, a shift in frequency causes a large change in output current linearly proportional to the frequency shift. Some comment on calibration, however, is necessary if this circuit is to be used as a transducer. The oscillation frequency f_0 is determined by the relation

$$f_0 = \frac{1}{2\pi \sqrt{LC}}$$

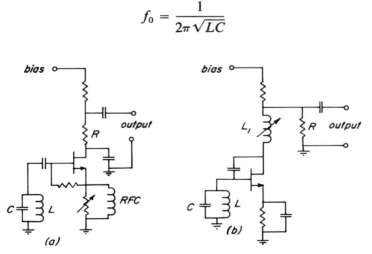

FIG. 8.1   Autodyne detector circuits. In (a) R_1 sets feedback for correct circuit operation; in (b) L_1 performs this function. The transducer variable element is either C or L and the LC combination is selected to produce oscillations at the frequency desired. The JFETs are selected according to the oscillation frequency desired. Output $= V_0$ and is developed by the IR drop across R.

Thus if the transducer is a variable capacitance device, for example, then

$$f_0 = \text{constant}/\sqrt{C}$$

This then shows that the output of the detector V_0 is proportional to $1/\sqrt{\Delta C}$ since $V_0 \sim \Delta f$. If the capacitance of the transducer varies linearly with excitation, then the detector output will vary as the inverse square-root of the excitation. To obtain a linear change in detector voltage for a linear change in excitation, either a nonlinear capacitor must be designed, or a nonlinear correction circuit using an operational amplifier may be used.

Linear operation can be approximated if the change in capacitance is small relative to the rest value of the capacitor. The analysis is as follows. Let f_0 be the center frequency, C_0 the rest value of the capacitor, ΔC the incremental change in capacitance, and L the inductance of the tuned circuit, so that

$$f_0 = \frac{1}{2\pi\sqrt{LC_0}}$$

$$\Delta f = \frac{1}{2\pi}\left[\frac{1}{\sqrt{L(C_0 + \Delta C)}} - \frac{1}{\sqrt{LC_0}}\right]$$

$$= \frac{1}{2\pi\sqrt{LC_0}}\left[\frac{1}{\sqrt{1 + \Delta C/C_0}} - 1\right]$$

If we recognize that $\Delta C/C_0 \ll 1$, then by the binomial expansion of the first two terms we obtain

$$\Delta f \sim \frac{1}{2\pi\sqrt{LC}}\left[1 - \Delta C/2C_0 - 1\right] = \frac{\Delta C}{4\pi C_0\sqrt{LC}}$$

Thus for *small* ΔC, $\Delta f \sim \Delta C$ and $\Delta V_0 \sim \Delta C$. Similar conditions and analyses apply for variable inductance devices.

8.1.2.   The Linear Discriminator Circuit

When the transducer variable capacitance or inductance forms part of the tuned circuit in a conventional oscillator, then it is appropriate to use a linear discriminator circuit to convert change in frequency to change in voltage. The idealized model and performance are shown in Fig. 8.2. For *small* deviations in frequency about a center frequency f_0, a simple tuned circuit, as shown in Fig. 8.3, may be used to achieve

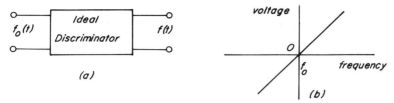

FIG. 8.2   Characteristics of ideal frequency discriminator.

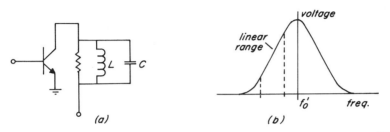

FIG. 8.3   A tuned circuit as a frequency discriminator: $2\pi f_0' = 1/\sqrt{LC}$.

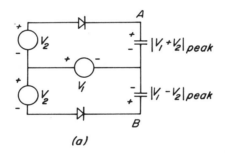

(a)

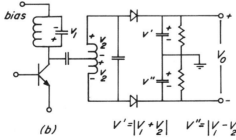

(b) $V' = |V_1 + V_2|$ $V'' = |V_1 - V_2|$

FIG. 8.4   Linear discriminator circuits: (a) ideal circuit: V_0 appears across $A - B$ and $V_0 = |V_1 + V_2| - |V_1 - V_2|$; (b) practical discriminator circuit, for which V_0 is defined as in (a).

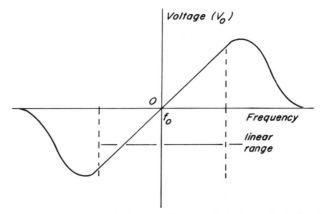

FIG. 8.5   Linear discriminator characteristic (practical circuit).

the characteristic shown in Fig. 8.2. To attain a "linear" characteristic, the tuned circuit is centered at a different frequency from f_0. The difficulty with this technique is that the variations in the desired signal voltage are superimposed upon the large nonzero "carrier" voltage produced by the oscillator. To achieve a practical device, a balanced system must be devised in which the carrier at frequency f_0 is removed. An ideal and practical system (the balanced discriminator) is shown in Figs. 8.4(a) and (b), respectively. The coupling into the discriminator is by transformer with a double-tuned circuit. The diode–capacitor combinations act as peak detectors. The voltage across the upper capacitor is $|V_1 + V_2|$, whereas the voltage across the lower capacitor is $|V_1 - V_2|$. The parallel RC combinations provide filtering so that the output signal V_0 is proportional to the instantaneous frequency of the input (oscillator) signal. Thus

$$V_0 = |V_1 + V_2| - |V_1 - V_2|$$

The practical discriminator has the characteristic shown in Fig. 8.5.

8.2. UNBALANCED-BRIDGE DETECTION SYSTEMS

For convenience, the equal-ratio arm bridge is normally used in transducer systems, as shown in Fig. 8.6. ΔR represents the change in resistance as a function of the associated transducer. When the bridge is balanced, $\Delta R = 0$. This is the initial or rest condition for the transducer. We will also assume that the impedance of the detector is very

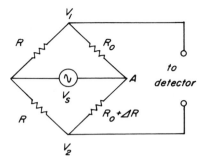

FIG. 8.6   Equal-ratio arm bridge circuit. Point A is the reference for the circuit equations cited in the text.

high (high-input-impedance amplifier or voltmeter) so that the current in the detector arm can be neglected. At balance (initial conditions for the transducer)

$$\Delta R = 0 \qquad V_1 = V_2$$

For unbalanced conditions (transducer activated, $\Delta R \neq 0$):

$$I_1 = V_s/(R + R_0)$$
$$V_1 = R_0 I_1 = R_0 V_s/(R + R_0)$$
$$I_2 = V_s/(R + R_0 + \Delta R)$$
$$V_2 = (R_0 + \Delta R)I_2 = (R_0 + \Delta R)V_s/(R + R_0 + \Delta R)$$

The unbalance voltage is $\Delta V = V_2 - V_1$

$$\Delta V = \frac{(R_0 + \Delta R)V_s}{R + R_0 + \Delta R} - \frac{R_0 V_s}{R + R_0}$$

$$= \frac{(R_0 + \Delta R)V_s}{(R + R_0)[1 + \Delta R/(R + R_0)]} - \frac{R_0 V_s}{R + R_0}$$

We now impose a constraint on the transducer that

$$\Delta R/(R + R_0) \ll 1$$

This condition is normally satisfied by strain gauges and thermistor devices used over a narrow temperature range. We can now write

$$V = \frac{(R_0 + \Delta R)V_s}{R + R_0} - \frac{R_0 V_s}{R + R_0}$$

$$= \frac{\Delta R V_s}{R + R_0}$$

Thus for small changes in resistance (ΔR), the bridge unbalance voltage is linearly proportional to the change in resistance. If ΔR is linearly proportional to transducer excitation, then ΔV is linearly proportional to excitation, and we have a linear system. Note, however, that the condition $\Delta R/(R + R_0) \ll 1$ must be satisfied.

Because ΔR is generally quite small, ΔV is also very small unless V_s is excessively large. For this reason, ΔV must be amplified before a useful output signal can be obtained. After amplification, the signal must be subjected to a detection technique to extract the signal envelope from the carrier waveform (bridge oscillator signal). In some systems, the bridge may be supplied from a battery. In this case, ΔV is a DC voltage which varies in proportion to ΔR. Amplification of small DC signals is difficult and it is better to use an AC "carrier" system in which the amplitude of the AC signal is proportional to changes in ΔR. Figure 8.7 illustrates a simple diode-detection circuit which can be used as an envelope detector.

The RC time constants for the circuit shown must be adjusted so that integration of the envelope signal does not occur. In order to obtain adequate filtering of the carrier and correct detection of the envelope, the frequency of the carrier (bridge oscillator) should be at least 100 times the maximum frequency associated with the transducer excitation signal. Typical carrier frequencies are in the range 1–10 kHz.

Generally only resistance bridges are used when the detector-arm offset-voltage technique is used. When AC signals are involved, the phase angle of V_1 and V_2 as well as amplitude is important. If reactive elements exist in the bridge, then phase angle differences can occur that render the offset-voltage technique invalid.

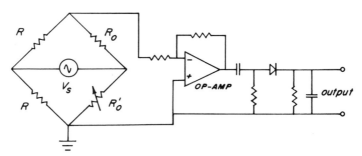

FIG. 8.7   Diode detector circuit for bridge: $R_0' = R_0 + \Delta R$. The resistance and capacitance of the RC combination following the diode must be adjusted such that the "carrier" is filtered out of the signal without distorting the envelope (modulating) signal.

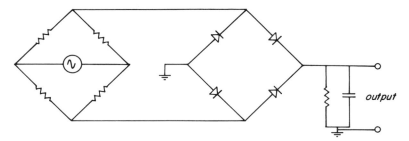

FIG. 8.8   Full-wave bridge detector circuit. The time constant of the RC combination across the output must be adjusted so that the "carrier" is filtered out of the signal without distorting the envelope (modulating) signal.

The detection system shown in Fig. 8.7 employs a single-ended (1 side ground) input signal. When a balanced signal is supplied either from a transformer or the alternate connection from a bridge that lacks the reference ground point, it may be desirable to use a full-wave bridge detector system, as shown in Fig. 8.8. The time constant requirement discussed above also applies here.

8.3.   IMPEDANCE MATCHING

It is frequently necessary to use rather long connecting cables between transducers and the amplifier or other electronic equipment to which they connect. This can introduce considerable shunt capacitance across the input of the amplifier because of the inherent capacitance per foot of shielded cable. Normally shielded cable is required to prevent pickup of spurious signals. If the shunt input resistance of the amplifier is high, this additional capacitance can generate a substantial RC time constant as indicated in Fig. 8.9. This time constant results in signal distortion and response delays.

Another problem is that most transducers are not ideal voltage sources (zero internal impedance) or ideal current sources (infinite internal impedance), but have internal impedance, as shown in

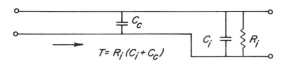

FIG. 8.9   Shunt cable capacitance and input circuitry.

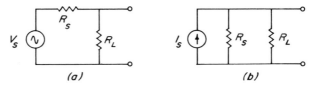

FIG. 8.10   Sources with source impedances: (a) voltage source; (b) current source. R_L is the external load on the source.

Fig. 8.10. If $R_L = R_s$, then half of the transducer voltage is lost, since

$$V_0 = R_L V_s/(R_L + R_s)$$

When $R_L < R_s$, this situation becomes worse.

We have defined two common problems associated with the use of transducers. These are shunt capacitance associated with connecting cables and internal resistance of the transducer source. Both produce signal degradation. A common solution consists of interposing a

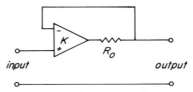

input   output

FIG. 8.11   Op-amp voltage follower circuit for impedance matching.

follower circuit between the transducer and the electronics to which it is connected. The basic principle is to reduce the impedance level, thereby reducing shunt capacitance effects and improving impedance matching. Either the operational amplifier voltage follower of Fig. 8.11 or a transistor emitter follower, as shown in Fig. 8.12, may be used.

The follower circuit presents a high input impedance to the output of the transducer, while at the same time producing a low output impedance that can be controlled. Thus we can design the follower so that its input impedance is $\gg R_s$. For this condition, then V_0 from the

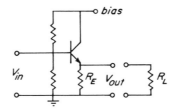

FIG. 8.12   Emitter follower circuit.

transducer will essentially equal V_s. The output impedance can be designed to match R_L for maximum power transfer, or to be $\ll R_L$ for maximum voltage transfer.

To solve the shunt capacitance problem, a low value of resistance is shunted across the amplifier input to reduce the time constant to an acceptable level. The follower is then used to match the transducer to the new R_L value.

The follower output impedance (for Fig. 8.11) is defined as follows:

$$R_{\text{out}} = R_0/K$$

where K is the open-loop gain of the op-amp. Thus if we have an op-amp with an open-loop gain of 500,000 and we want an output impedance of 10 Ω, we can find what value of R_0 to use from

$$R_0 = KR_{\text{out}} = (500,000)(10) = 5 \text{ M}\Omega$$

For Fig. 8.12,

$$R_0 = \left[\frac{1}{R_E} + h_{0e} + \frac{h_{fe} + 1}{h_{ie} + R_s} \right]^{-1}$$

where the h's are the respective h-parameters of the transistor used and R_s is the impedance of the transducer. If R_s is small, then

$$R_0 \sim 0.0259/I_E$$

where I_E is the DC bias current, and the shunting effect of the biasing resistors has been neglected.

The input impedance for the op-amp follower circuit is essentially the input impedance for the op-amp alone, which is very high. The input impedance for the transistor circuit is given by

$$R_i \sim h_{ie} + (h_{fe} + 1)R_L'$$

where $R_L' = R_L R_E/(R_L + R_E)$. Some typical transistor values are: $h_{ie} = 4 \text{ k}\Omega$; $h_{0e} = 2.5 \times 10^{-5}$; $R_E = 2 \text{ k}\Omega$; $h_{fe} = 250 \Omega$; $h_{re} = 4 \times 10^{-4}$; $R_L = 2 \text{ k}\Omega$; $I_E = 2 \text{ mA}$; $R_S = 100 \Omega$; $R_i \sim 4000 + (250 + 1) \times R_L' = 4251 R_L'$.

If $R_L' = 1 \text{ k}\Omega$, then $R_i \sim 4.25 \text{ M}\Omega$

$$R_0 = \left[0.0005 + 2.5 \times 10^{-5} + \frac{250 + 1}{4000 + 100} \right]^{-1} = 16.2 \ \Omega$$

From the approximation

$$R_0 \sim 0.0259/I_E = 0.0259/0.002 = 12.95 \ \Omega$$

Thus the voltage follower solves both the time constant problem generated by the shunt capacitance and the impedance matching problems.

8.4.   REFERENCES

Alley, C. L., and K. W. Atwood, 1973, *Electronic Engineering*, 3rd Ed., Wiley, New York.
Brophy, J. J., 1977, *Basic Electronics for Scientists*, 3rd Ed., McGraw-Hill, New York.
Malmstadt, H. V., C. G. Enke, and S. R. Crouch, 1974, *Electronic Measurements for Scientists*, Benjamin, Menlo Park, California.

Section 4

RECORDING AND DISPLAY

9

Display and Recording Systems

A necessary component of any instrumentation system is a mechanism for displaying the signals of interest or the data collected. One may wish to have, in some cases, immediate indication of system response or, in other instances, a permanent record for subsequent analysis. Thus a selection must be made from available devices such as LED displays, oscilloscopes, chart recorders, and tape systems.

The response time of display and recording devices is of primary importance in the selection of the appropriate instrument to use in an instrumentation system. Inadequate response time can result in signal distortion and loss of data. When one is working with low-frequency (on the order of 100 Hz) analog signals, oscilloscopes, paper chart recorders, and some tape systems are appropriate. Oscilloscopes are not appropriate for signals which vary very slowly in time, unless a storage scope is used. Conventional direct-recording tape systems are also inappropriate, although FM systems can be used. Paper chart recorders are perhaps the best suited for this application. For high frequencies (1 kHz and above), paper chart recorders cannot respond adequately because of the inertia inherent in electromechanical recording systems. Tape and oscilloscope systems are suited to high-frequency applications.

In addition to the response time of the recording instrument, one must also be concerned with the input impedance of the device. As illustrated in Chapter 8, the combination of cable capacitance and a high shunt resistive component in an input circuit can produce substantial time constants. Thus impedance matching networks may be required to couple properly into a recording or display device.

Other than the problem cited in the paragraph above, no discussion of conventional oscilloscopes is required. These should be familiar instruments to anyone who has had any experience with electronic equipment. A storage scope generally proves more useful in biomedical work than a conventional model because it is more versatile.

9.1. CHART RECORDERS AND PLOTTERS

Chart recorders are either linear or X–Y plotters. In the linear or strip-chart recorder, a paper chart is driven at constant rate by a drive motor. The tracing is made by a pen that moves in a direction normal to the chart motion. In this manner, a record of some variable (the transducer input) as a function of time is obtained. The pen may operate based upon two principles. In some new instruments, a sensitive motor is the transducer. The pen assembly is attached to the motor shaft and the motor is excited by an electrical signal. The motor may be connected in either the series or shunt excitation modes.

Normally a signal processing amplifier, as shown in Fig. 9.1, does not produce sufficient power to drive such a motor. This means that a separate motor-driving power amplifier must be constructed. The advantage of this system is ruggedness and its adaptability to use in portable equipment. The disadvantage is the extra power amplifier required and its current drain when the recorder must be operated from batteries. Generally the motors are small and the overall recorder is light in weight.

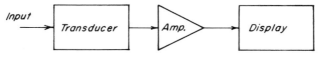

FIG. 9.1   Basic instrumentation system.

The more conventional strip-chart recorder uses a d'Arsonval type of meter movement in which the usual meter pointer is replaced by the recording pen assembly. In order to achieve reasonable sensitivity and compliance of the movement, quite large magnets are required. These cause the instruments to be heavy and less adapted to portable use in small equipment.

X–Y plotters that plot one variable against another variable work on a slightly different principle. Two servomotors are used, and there are two independent self-balancing servomechanism systems. One servomechanism moves the pen carriage in the vertical direction (y coordinate), while the other moves the entire carriage across the chart in the horizontal direction (x coordinate). The carriage holds the pen assembly. A drive mechanism permits the pen assembly to travel the length of the carriage. Both the X and Y servomechanisms respond to input signals applied to their respective input terminals. The relative motion between the two servomechanisms that results from the applied signals produces a Cartesian coordinate graph. Figure 9.2 shows a system block diagram for one coordinate axis. The servomotors are connected to drive strings, which actually move the pen and carriage.

Graphic recorders that use paper charts have a number of limitations. The major one is frequency response. Because of the mechanical inertia associated with these devices, pen response is slow. About 1 kHz is the maximum sinusoidal signal that can be recorded and 100 Hz is more normal. Linear strip chart recorders have an additional limitation because of the necessarily fixed location of the pen. The pen must pivot about a shift which creates a radius relative to the tip. This means that the pen produces curved vertical lines rather than straight lines. This

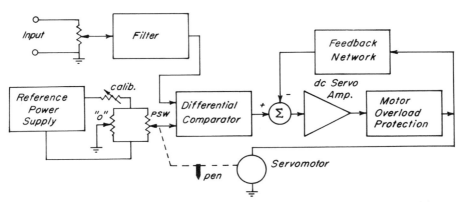

FIG. 9.2   Block diagram for a single channel in an X–Y plotter (or a channel of some linear strip-chart recorders). PSW is a precision slidewire.

effect is minimized by making the pen holder long relative to the arc length through which it moves. There are limitations, however, on the length of the holder relative to mechanical stability and inertia. This condition produces a certain amount of waveform distortion.

There are problems associated with the pen. Both dry and wet pens are used. The wet pen consists of a capillary tube connected to an ink reservoir. Associated problems are plugging of the tube and smearing of the ink record before the ink dries. Periods of disuse cause the ink to dry in the reservoir and low-temperature conditions result in the ink freezing.

To rectify the disadvantages of wet recording, several dry techniques have been developed. One uses a temperature-sensitive paper and replaces the capillary tube by a heated stylus. There are several drawbacks to this method. The stylus heater draws a large current, which presents a problem with battery drain in portable equipment. Changes in stylus temperature cause irregularities in the recorded tracing and the stylus heaters burn out relatively easily. The recording paper is sensitive to heat and many volatile chemicals which sometimes makes storage difficult.

One method that solves the temperature problems utilizes a pressure or chemically sensitive paper and a metal stylus, frequently of brass. Contact of the stylus on the paper produces a chemical reaction which results in discoloration of the paper. The tracings tend to be light and the paper is chemically reactive, which again presents storage difficulties.

Several designs have been developed to overcome the high-frequency limitation associated with paper-chart-and-pen recorders. One method employs a sensitive galvanometer in which a small mirror replaces the pen assembly. A light is focused on the mirror and is in turn reflected upon light-sensitive paper. Usually a light source high in ultraviolet components is used. This system eliminates the inertia and friction problems associated with pen recorders. Deflection of the mirror by current in the galvanometer coil causes the light beam to "write" on the paper. Photographic film may be used in place of the paper. The disadvantage of this system is its photosensitive nature. Recordings must be made in subdued light or total darkness when film is used. Films must be photographically developed and there is a storage problem when the UV-sensitive paper is used. Sunlight or light from fluorescent lights causes it to darken and the trace is lost.

Another approach to high-frequency recording is the ink-jet recorder. In this system a very small high-pressure nozzle is attached to the galvanometer coil. A jet of ink is sprayed on the recording paper.

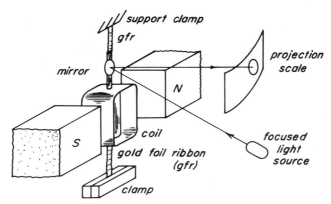

FIG. 9.3   Rotating mirror light-beam recorder movement.

There is a certain amount of damping in this system owing to the hose that supplies the nozzle. Even with these techniques, about 10 kHz is the maximum frequency that can be recorded. Figures 9.3 and 9.4 illustrate these methods.

With any paper chart system, one must be concerned with response time, input impedance to the device, and sensitivity. There is little that can be done about the response time since this relates to the electromechanical properties of the recorder mechanism. Input impedance problems can usually be solved using an op-amp follower circuit of the type described in Chapter 8. Lack of sensitivity can usually be rectified by using a preamplifier ahead of the recorder.

9.2. TAPE RECORDING DEVICES

A variety of tape recording devices is available for recording both analog and digital signals. The choice of recording technique depends

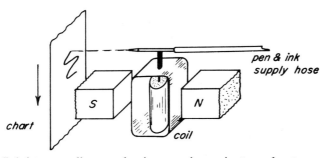

FIG. 9.4   Ink-jet recording mechanism; various pivots and return springs are not shown in the diagram. See also Offner, 1967.

upon what sort of subsequent data analysis is anticipated. If one wishes simply to have a permanent analog record for display later on an oscilloscope or chart recorder, then direct recording or FM recording is indicated. On the other hand, if the data are to be subjected to extensive statistical or frequency spectrum analysis, then one should consider analog-to-digital conversion of the data and recording digitally in a format that can be processed by a computer.

The selection of the correct analog tape recording method depends upon the frequency spectrum of the signal to be recorded. With the exception of certain specialized recorders, the majority of commercially available analog units respond over the audio spectrum. Generally speaking, response is poor at frequencies below 50 Hz and above 15 kHz. This means that conventional recorders are not applicable when the signals to be recorded have substantial low-frequency components or a DC level. They are equally inapplicable for recording signals with significant harmonic components above 15–20 kHz.

The basic problem is not in the electronics, but rather in the design of recording and playback heads used to "write" and "read" magnetic tape. The construction of typical recording heads is shown in Fig. 9.5, and Fig. 9.6 shows a playback head and its equivalent magnetic circuit. It is difficult to construct a magnetic circuit that will respond over a wide frequency spectrum (essentially equivalent to a low-pass characteristic). The low-pass characteristic is required if one is to record from DC to some upper frequency limit. Basically the magnetic circuit of a tape head represents a band-pass filter with the characteristic shown in Fig. 9.7. Thus recorder heads can be designed

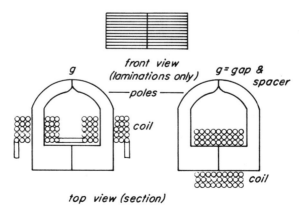

FIG. 9.5 Fabrication of magnetic recording heads. Heads shown are the modified-ring type. See also Stewart, 1958.

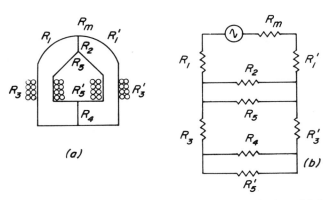

FIG. 9.6 Construction of (a) and magnetic network equivalent (b) for magnetic reproducing head: R_m = magnetic reluctance of the recording medium (tape) in contact with the head; R_1, R_1' = reluctances of tape-head contact surfaces; R_2, R_4 = the front and rear gap reluctances, respectively; R_3, R_3' = reluctances of the core halves; R_5, R_5' = leakage reluctances at the front and rear gaps respectively. See also Stewart, 1958.

to operate about some center frequency (within limits), but they cannot be designed to operate at DC or at very low frequencies.

For signals whose major spectral components lie outside of the range 50–15,000 Hz, two recording choices exist. An FM tape system can be used, or one can A/D convert the signal and record the information digitally. The FM system operates on the same principle as an FM radio. A suitable carrier frequency is selected and the signal to be recorded is used to frequency modulate the carrier about its center value. This technique solves the problem of low frequency and DC components. It is not applicable to the recording of waveforms whose major spectral components are greater than 15 kHz. To recover the analog signal recorded by the FM technique, the reproduction channel includes an

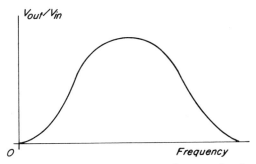

FIG. 9.7   Frequency response for tape head.

FM demodulator, such as the discriminator circuit shown in Chapter 8. The frequency response of an FM tape recorder depends upon the rate at which the tape is driven past the record head. Table 9.1 indicates typical machine characteristics.

Table 9.1
FM Tape Recorder Characteristics

Tape speed, in./s	Carrier frequency, kHz	Deviation, kHz	Frequency response, kHz
$1\frac{7}{8}$	1.6875	2.363–1.012	DC to 0.313
$3\frac{3}{4}$	3.375	4.725–2.025	DC to 0.625
$7\frac{1}{2}$	6.75	9.450–4.050	DC to 1.250
$7\frac{1}{2}$	13.5	18.9–8.1	DC to 2.50
$7\frac{1}{2}$	27.0	37.8–16.2	DC to 5.0

When high frequencies must be recorded (above 15 kHz), the usual technique is digital recording. The analog signal is processed in an analog-to-digital converter and the information stored as discrete voltage levels on magnetic tape. In order to recover the analog signal for display on an oscilloscope or paper chart recorder, the information stored on magnetic tape must be passed through a digital-to-analog converter. An alternative to tape recording is disc recording. The high-speed disc units associated with large-scale computers are quite expensive, but portable equipment that uses "floppy discs" is comparable in cost to conventional tape units.

A major advantage of digital recording is that the data can be stored on tapes or discs in formats that are easily handled by a computer. This eliminates initial storage of the information on analog tape, with subsequent conversion to digital tape for computer processing.

9.2.1. Tape Configurations

Much analog recording is accomplished using the conventional reel-to-reel method. When large amounts of data must be recorded for which signal fidelity must be preserved, this is probably the best technique. Oversize reels can be run at speeds of 15 or 7.5 in./s. When the high-frequency response requirements are not exacting, one should consider the use of tape cassettes. These are smaller, lighter, and much more portable than reels. Cassettes are also being used extensively for digital recording, especially in minicomputer applications. For given amounts of data recording capacity, cassette tape transports are considerably smaller than reel-to-reel machines.

9.2.2.   Tape Transport Considerations

A large variety of analog tape transports is available. In selecting a unit, cost alone should not be the measure for selection. Many of the high priced units include a number of convenience features which may not be useful to, or worse, even interfere with, data recording. The basic tape record/reproduction heads and transport mechanism may be no better than or even inferior to lower priced units that incorporate fewer conveniences.

The basic features to check are mechanical specifications, electrical specifications, and versatility. In a reel-to-reel machine, one should check the number of motors. Is there a single motor which drives the capstan and reels through belts and pulleys, or is there a separate capstan motor with reel torque motors? The latter is a more stable arrangement. Is the capstan motor synchronous or induction? In most cases, the synchronous motor is desirable to assure speed regulation. What are the flutter and wow specifications for the transport? Basically flutter is high frequency variation or undulation of the recorded signal resulting from rapid speed variations of the transport. This is frequently related to motor speed variations. Wow is low-frequency modulation of the recorded signal. It is easily heard when sustained musical notes (trumpet and piano in particular) are reproduced. Wow may be motor related, but is usually associated with changing drive-belt tensions and other mechanical malfunctions in the transport mechanism. All transports exhibit some wow and flutter. Excessive amounts are indicative of poor design or poor components, or both.

Another mechanical feature that should be checked is the mechanism by which the tape is held in contact with the heads. In some machines friction is provided by posts located on either side of the head assembly, as shown in Fig. 9.8. In this system, uneven reel motion because

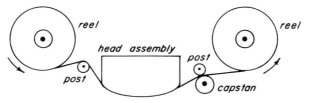

FIG. 9.8   Reel-to-reel tape drive assembly.

of sticking tape can cause the tape to move away from the heads with resultant distortion or loss of signal. Many machines use a pressure pad that keeps the tape in contact with the head assembly at constant pressure. The disadvantage of this system is increased tape and head wear.

The selector switches that form the functional controls for the transport should be carefully checked. Both their reliability and ability to function positively must be examined. This is frequently the weak point in many transports. The overall construction of the transport is also important. Durability should be inspected.

Although there is wide divergence in the design and construction of reel-to-reel transports, cassette decks are generally quite similar. A main consideration here is ease in servicing. Cassette recorders appear generally to have higher maintenance requirements than reel-to-reel machines. Some items to be examined are ease of cleaning heads and drive mechanisms, possible drive-belt slippage problems, and the load/eject mechanism.

The electrical specifications of greatest importance are frequency response, signal-to-noise level, interchannel interference in multichannel machines, and tape drive speeds in reel-to-reel machines. As noted above, tape heads are essentially band-pass filters. The overall frequency response of a tape recorder depends upon both the heads themselves and the amplifiers which are associated with them. Figure 9.9 illustrates a frequency response characteristic for a reproducing head, the amplifier characteristic, and the resultant channel characteristic for the head–amplifier combination. Amplifiers are designed to compensate for the poor frequency response of the heads in both the

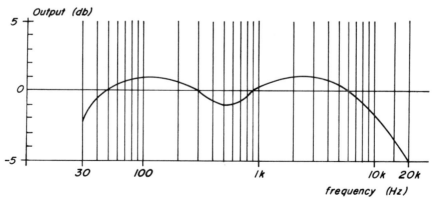

FIG. 9.9   Frequency response for tape recorder reproducing channel.

record and reproduce channels. In the record mode, this is usually called preemphasis and in the reproduce mode, playback equalization. The frequency characteristic of the head determines how much compensation can be applied by the amplifier. There is both a low- and high-frequency cutoff beyond which compensation is inapplicable.

Signal-to-noise ratio (S/N) is important for signal fidelity and clarity, particularly at low recording levels. There is always noise generated by the passage of the tape over the heads. This is a function of friction between the heads and the iron oxide particles on the tape. The important area is the S/N associated with the heads and amplifiers. Preemphasis and equalization networks, if improperly designed, can generate significant electrical noise.

Interchannel interference relates to poor layout of circuit components and poor head design. Channel separation should be better than 45 dB in a correctly designed instrument.

Cassette decks operate at a fixed speed. Since tape speed and frequency response are directly related, the only mechanism for improving response in such units is through head design and amplifier compensation. S/N is also directly related to tape speed and is the reason for the Dolby circuits in modern cassette decks.

A good reel-to-reel machine will have provision for several tape speeds, usually $1\frac{7}{8}$, $3\frac{3}{4}$, $7\frac{1}{2}$, and 15 in./s. The highest speed produces the widest frequency response and the lowest S/N.

Harmonic distortion (HMD) and intermodulation distortion (IMD) are also important. The major source is usually the head, because it is a magnetic circuit with some nonlinear properties. Harmonic distortion is the production of frequency components not present in the original signal. These components are integer multiples of the original frequency components. Interaction of two frequencies with the production of their sum and difference frequencies and multiples is intermodulation distortion. Saturation and other nonlinear properties of both heads and amplifiers can produce both HMD and IMD. For good design, generally speaking HMD $< 2\%$, IMD $< 0.5\%$.

Machine versatility is important if one intends to modify or expand an instrumentation system. Because of the nature of the cassette mechanism, little modification can be done, though reel-to-reel machines, properly designed, have the potential for substantial modification. Items to investigate are the ability to increase reel size for increased data-handling capacity, and blank spaces in the head assembly for the installation of additional heads.

Generally speaking, it is better to use separate heads for erase, record, and playback. Combination playback/record heads are a compromise and cannot provide the fidelity of reproduction of separate heads. It must be recognized that since tape heads are magnetic circuits, they are subject to various nonlinear phenomena. The basic materials used obey a *B–H* field hysteresis loop, as shown in Figure 2.2. In normal operation, the magnetic material is not permanently magnetized and operation is along the reversible first magnetization curve. During use, the tape heads become magnetized to a certain extent as a result of the tape moving across the heads and the signals applied to them. This results in a decrease in S/N and degradation in signal reproduction. To avoid this, the heads must be periodically demagnetized.

Tape transports require regular maintenance, which includes lubrication of appropriate mechanical parts, head cleaning and demagnetization, cleaning of the capstan, drive, and idler pulleys, and readjustment of head alignment as required. Cassette decks are particularly susceptible to dust and dirt pickup and require cleaning on a regular schedule. They also tend to destroy tapes. This usually results from differential tensions in the capstan drive assemblies. The cause of this is normally dirt or grease on the drive belts or pulleys and loss of tension in retaining springs. When cassette machines are serviced, it is essential that all of the moving parts associated with tape cassette operation be *thoroughly* cleaned with ethyl alcohol or a solvent specifically designed for this application.

9.3.   ALPHANUMERIC DISPLAYS

Many instruments are designed to provide a direct numerical readout, or in some instances, an alphabetical display. In these cases, the signals are processed digitally and interpreted by logic circuits as numbers or letters. The logic circuits then direct appropriate driving voltages or currents to the display elements.

Displays may be incandescent elements, such as the RCA Numitron, gas discharge devices (Burroughs Nixie® tubes and Sperry planar displays), LED alphanumeric matrices, and liquid crystal elements. Displays of this nature do not produce any permanent records, although some instruments are designed such that the display signal is also available as a digital output for recording on magnetic tape. Displays may be single or multicharacter, in that only one numeral or letter is displayed at a time, or a sequence of alphanumeric characters is displayed simultaneously.

The RCA Numitron is a single character display composed of seven incandescent segments mounted on a planar substrate. A decimal point is also provided. The unit is housed in an evacuated 9-pin miniature vacuum-tube envelope. By applying DC or AC voltages to the appropriate pins, the numerals 0–9 and a decimal point are obtained. The advantage of this type of display is its brightness, so that it can be viewed under high ambient light conditions. This type of display has two disadvantages: high current requirement for the segment filaments, and potential burnout of segments.

Various gas discharge display devices are available. The "old standby" is the Burroughs Nixie® tube first introduced in the 1950s. It consists of a stacked array of planar elements, each of which is a numeral. Each of the numerals is selectively the cathode for a gas discharge diode. The display array is housed in a neon–argon gas-filled glass envelope which is viewed from the top. The numerals are illuminated by the cathode-glow discharge that takes place when a particular numeral element is energized. National Electronics makes a similar device that is viewed from the side.

Gas discharge displays of this nature have several disadvantages: limited lifetime as a result of gas-cleanup and sputtering; and high excitation voltages (> 90 V DC). They do produce a fairly bright light and the current drain is less than incandescent displays.

Recently, Burroughs Corporation has introduced a "Self Scan" display that can simultaneously exhibit 16–32 characters, depending upon model, including letters, numerals, and punctuation. The maximum character size is about 0.4 in. It is a gas discharge array which requires 250 V DC at currents up to 100 mA depending upon the model. A logic package comes with the display which accepts the Standard ASCII teletype code. The display moves sequentially as each new character is entered, and with each new character entered at one end, an existing character is removed. The unit is planar and has the approximate dimensions of $1.5'' \times 10''$ for the display face.

Another type of gaseous display, also a planar design, is marketed by Sperry Rand. The characters are formed of gas-filled tubes and operate along the lines of a neon sign.

A popular display is the LED matrix, as shown in Fig. 9.10. This is a 5×7 character alphanumeric unit manufactured by Texas Instruments, Inc. The display characters and logic coding are shown in Fig. 9.11. Figure 9.12 illustrates the auxiliary circuitry required. The TMS 4100 shown is a decoder driver. The display requires a maximum of 400 mA and the center wavelength is 660 nm, and character height is 0.34 in. The individual units may be mounted side-by-side on 0.45 in. centers to construct multiple displays.

Another display technique which is increasing in popularity uses

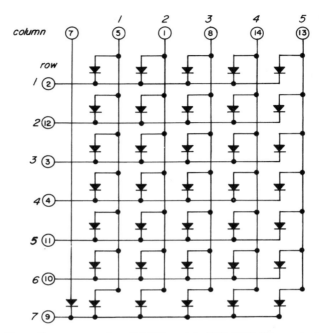

FIG. 9.10   Texas Instruments 5 × 7 LED matrix (TIL 305). The encircled numbers are the DIP-device pin connections.

liquid crystals (LCDs). Liquid crystals are cholesterol esters whose optical characteristics change as a function of temperature. The devices are passive in that they do not emit light in themselves, but scatter incident light. They can be used in either the transmissive or reflective mode to produce light of any color. Thus they can be observed under daylight or artificial light conditions. Individual characters are available in heights up to 1 in. As in the RCA Numitron, the characters are normally composed of seven segments. They are available in units ranging from one to eight characters in a planar configuration; 5 × 7 matrix displays are also available. Typical operating characteristics are:

Voltage, 3–40 V
Frequency, 30–1000 Hz
Temperature, −20 to +70°C
Current consumption, up to 30 μA

Additional logic decoder-driver units are required to drive these displays. LEDs may be driven from either AC or DC sources. LCDs,

TYPICAL APPLICATION DATA
RESULTANT DISPLAYS
USING TMS4103JC OR TMS4103NC
WITH USASCII CODED INPUTS

FIG. 9.11   Coding and characters in Texas Instruments TIL 305 5 × 7 alpha-numeric LED display. Positive logic: 1 = H = 2–5.5 V; 0 = L = 0–0.8 V. Reproduced with permission from *The Optoelectronics Data Book for Design Engineers*, published by Texas Instruments Incorporated, Dallas, TX.

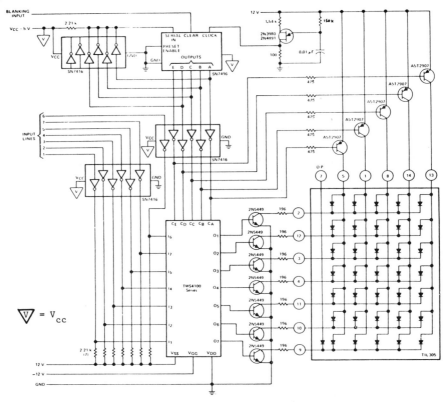

FIG. 9.12 Applications diagram for Texas Instruments TIL 305 5 × 7 LED display. Reproduced with permission from *The Optoelectronics Data Book for Design Engineers*, published by Texas Instruments Incorporated, Dallas, TX.

however, must be operated so that there is no DC component across the crystal. Typically the back plane of the display is driven by a 30–300 Hz square wave. LCDs have the advantage of very low current drain (μA), while LED displays are high current drain (in tens to hundreds of mA, depending upon the display and number of elements). In some applications, LED displays are more visible than LCD displays.

9.4. SUMMARY

As we have seen from the foregoing, there are a number of available devices for recording and displaying the data collected by an instrumentation system. Any one or several of these units may be employed.

Specific application, convenience, and cost normally dictate the selection. New devices are regularly appearing on the market.

9.5.   REFERENCES

Offner, F. F., 1967, *Electronics for Biologists*, McGraw-Hill, New York.
Stewart, W. E., 1958, *Magnetic Recording Techniques*, McGraw-Hill, New York.
Specification sheets from: Burroughs Corp., Data Inc., Hewlett Packard Co., MFE Corp., Radio Corporation of America, Texas Instruments, Inc.

Instrumentation Systems

In this chapter, we will examine specific types of instrumentation systems used in environmental, clinical, and rehabilitation applications. It is not possible to discuss all types of instruments, since an encyclopedia would be required. The examples selected for review in this chapter were chosen to illustrate transducer techniques discussed previously.

10.1. SPECTROPHOTOMETERS

The spectrophotometer is an instrument with many applications. The instrument itself simply measures optical density, or percent transmission of light through a liquid sample. This information may be used to determine the concentration of organic and inorganic compounds in solution. Enzyme reactions and other biochemical studies may be carried out using these instruments. The basic system is shown in block diagram form in Fig. 10.1. The basic components are: light source, collimator, optical line separator, sample chamber, optical detector, amplifier, and readout. There are several critical design considerations that will now be outlined.

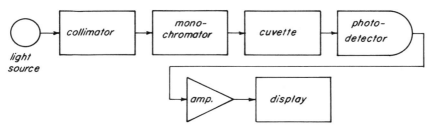

FIG. 10.1   Block diagram of spectrophotometer system.

The major consideration is the inherent nonlinear nature of the measurement. The amount of light transmitted through a sample obeys Beer's law, which is an exponential relation. It is usually formulated as

$$I(\alpha, x) = I_0 e^{-c\alpha x}$$

where $I(\alpha, x)$ is the intensity of light transmitted through the sample as a function of sample concentration and/or sample thickness, I_0 is the initial intensity of the light striking the sample, x is the sample thickness, α is the concentration of substance in solution, and c is a conversion constant to provide consistent units. Normally in a spectrophotometer x is fixed (the sample cuvettes are of standard size) and α is the quantity that varies. Thus if the sample thickness is a fixed distance $x = \tau$, then

$$I(\alpha) = I_0 e^{-c\alpha\tau} = I_0 e^{-c_0\alpha}$$

where $c_0 = c\tau$. Thus the light available to the optical detector is proportional to sample absorption (concentration) according to an exponential law. Most optical detectors operate as photoconductive devices, which means that device current is linearly proportional to impinging light intensity. If the optical detector is connected in the commonly used configuration shown in Fig. 10.2, then the voltage V

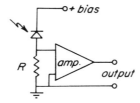

FIG. 10.2   Reverse-biased photodiode. R is generally quite large (10^5–10^6 Ω range).

presented to the amplifier shown in Fig. 10.1 varies with respect to concentration α as follows:

$$V = RI$$
$$I = KI(\alpha) = KI_0 e^{-c_0\alpha}$$

Thus

$$V = RKI_0 e^{-c_0\alpha}$$

where K is a conversion constant for the optical characteristics of the photodetector.

If a conventional linear amplifier is used, then its output voltage/ current will be exponentially related to sample concentration. This means that the display device will have a nonlinear calibration. For this reason, one usually manipulates Beer's Law in the following manner:

$$\frac{I(\alpha)}{I_0} = e^{-c_0\alpha}$$

$$\ln \frac{I(\alpha)}{I_0} = -c_0\alpha$$

$$\log \frac{I(\alpha)}{I_0} = -c_0'\alpha$$

where c_0' incorporates the ln to log conversion factor. $I(\alpha)/I_0$ is defined as the transmittance of the sample and is usually expressed in percent. Thus

$$\frac{I(\alpha)}{I_0} \times 100 = \text{percent transmission} \,(\%T)$$

or

$$e^{-c_0'\alpha} \times 100 = \%T$$

If we take the common logarithm of both sides of the above relation, then

$$\log \frac{I(\alpha)}{I_0} + \log 100 = \log(\%T)$$

But

$$\log I(\alpha)/I_0 = -c_0'\alpha$$

Thus

$$-c_0'\alpha + 2 = \log(\%T)$$

or

$$2 - \log(\%T) = c_0'\alpha$$

The quantity $2 - \log(\%T)$ is known as the absorbance A or the optical density (OD) of the sample. The absorbance is directly proportional to concentration. Absorbance ranges in value from 0 to ∞, and $\%T$ ranges from 100 to 0%. Beer's Law is exact only for monochromatic light.

Since $V = RKI_0\,e^{-c_0\alpha}$, the voltage as a function of percent transmission is

$$V = RK'\,e^{-c_0\alpha}$$

If the instrument is to display $\%T$ on a linear scale, as is the case with many spectrophotometers, then the amplifier, as shown in Fig. 10.1, must be a logarithmic amplifier, such that $V_{out} = \ln V_{in}$.[†] To avoid a DC baseline offset, V is normalized such that

$$V_{in} = \frac{V}{RK'} = e^{-c_0\alpha}$$

then

$$V_{out} = -c_0\alpha$$

and the instrument readout is calibrated in linear $\%T$ with $V_{out} \sim \alpha$.

Hence we see that the amplifier used in a spectrophotometer must be logarithmic in function if a linear readout proportional to concentration is to be obtained. A nonlinear (logarithmic) absorbance scale from 0 to 1 is usually included. Initial calibration of spectrophotometers is critical.

The light source used in these instruments is also critical. Normally an incandescent tungsten filament is used to span the infrared (IR) and visible wavelengths. A mercury vapor or deuterium lamp is usually used in the ultraviolet (UV) range for wavelengths down to 200 nm. When a tungsten source is used, it must be stabilized in terms of electrical current supplied, since light intensity varies as the fourth power of applied current I. If the filament has a resistance R, then the temperature of the filament is $T = KRI^2$, where K is a proportionality constant. The energy emitted by the filament is proportional to T^4 with intensity proportional to T^2. Thus $I \sim (KRI^2)^2 \sim K'I^4$. Wavelength λ

[†] Some photovoltaic devices exhibit a logarithmic response as a function of incident light intensity. In a few cases when such devices are used, the logarithmic characteristic of the device corrects for the exponential behavior of the signal source, and a linear amplifier may then be used.

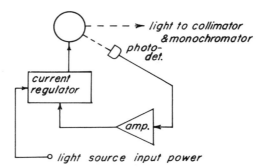

FIG. 10.3   Servosystem for regulating light source current and intensity.

is inversely proportional to temperature ($\lambda \sim 1/T$). Gas discharge lamps such as Hg vapor sources are inherently unstable with regard to amount of light output. This problem can be solved by use of a split beam arrangement, as discussed subsequently.

Incandescent light sources can be regulated by the system shown in Fig. 10.3. This is a feedback system in which the light output from the lamp is continuously monitored by a photodiode. Any change in light intensity causes the diode current to vary. The diode in turn controls a current regulator in series with the lamp, and constant light output is obtained.

The split beam technique may also be used, as shown in Fig. 10.4. It is most easily accomplished using half-silvered mirrors that reflect and transmit equal amounts of light. The light from the two channels is received by the photodetectors and the ratio of their outputs is determined. This is the quantity displayed and light fluctuations are automatically compensated. The ratio is actually achieved by a difference amplifier, as shown in Fig. 10.4.

The disadvantage of this system is that half of the light intensity available to the sample cuvette from the light source is lost. Half-

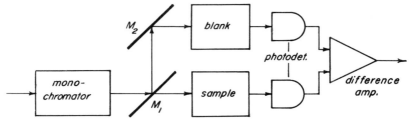

FIG. 10.4   Beam splitting arrangement for spectrophotometer: mirror M_1 is half-silvered; M_2 is fully silvered.

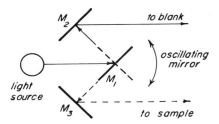

FIG. 10.5   Alternative beam splitting arrangement using a vibrating or oscillating mirror.

silvered mirrors may be made by two techniques. One involves depositing a thin layer of silver on a glass substrate such that 50% transmission and 50% reflection of light is achieved. In the second technique, the silver is more heavily deposited in a dot matrix, such that 50% of the mirror area is covered with dots. To avoid the 50% light loss, a fully-silvered vibrating mirror can be used as shown schematically in Figure 10.5. Although full light intensity is now available to the sample cuvette, the comparison of the two light signals becomes more difficult as they are not coincident in time. This method effectively produces a chopped light source which does have some advantages, as discussed below.

The photodetectors used in spectrophotometers must be selected to provide a uniform response over the range of operation of the unit. Since photodetectors are wavelength sensitive, usually several units with different spectral characteristics must be used to cover the entire spectrum from IR to UV. Dark current may also be a problem. This is the current output from a photodetector when no light strikes it. If excessive, it can produce an offset in the $\%T$ or A readings. To compensate for this, a chopped light source can be used. Normally a disc, as shown in Figure 10.6, is rotated in front of the light source. The disc provides two functions by chopping the light. It allows comparison of the light current against the dark current and also provides an AC

FIG. 10.6   Light chopping disc for spectrophotometer.

signal for amplification. Thus baseline drift associated with DC amplifiers is eliminated.

Many spectrophotometers are of the manual type. That is, the instrument must be adjusted for each wavelength selected. One uses a blank cuvette filled with the sample solvent and adjusts the instrument for $100\% T$ or $0 A$. In effect, this is manual correction for the spectral response characteristics of the photodetector. Scanning spectrophotometers are also available. There are several ways in which compensation for the spectral characteristics of the detectors may be obtained. Arrays of photodetectors may be connected in parallel to provide a nearly flat spectral response. This method is expensive and requires splitting of the light beam so that each detector receives an equal amount of light. A second method involves running a calibration curve using the light source and detector referred to a standard light source. The spectral characteristic of the system is then stored in a special purpose computer programmed to perform the correction automatically. A third, and simple, method utilizes a set of filters in the light path. These are designed to "smooth out" the source and detector characteristics so that the detector output voltage is essentially constant over the full spectral range of the instrument as long as the light striking the detector is of constant intensity.

Two types of monochromators are in general use. The simplest and cheapest is a prism to split out the spectral lines from the light source. This is used in conjunction with mechanical entrance and exit slits, as shown in Fig. 10.7, so that narrow bands of wavelengths may be selected. It is not possible to reduce the opening of the exit slit such that a single spectral line is selected. Below a certain slit width, diffraction occurs. A better monochromator method utilizes a diffraction grating. Usually replica gratings are used because of the very high cost of primary gratings. Good-quality replica gratings are available in several rulings-per-inch categories. Typical values range from 15,000 to 20,000 rulings/in. Primary gratings are normally metal. Replica gratings are made by coating the primary with collodion.

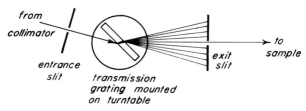

FIG. 10.7   Transmission grating monochromator.

When dry, the collodion replica is stripped from the "mold" and is mounted on a glass substrate. The higher the number of rulings per inch, the better is the resolution of the grating. Gratings may be either reflection or transmission types.

A number of other factors must be considered in spectrophotometer design. Among these are length of light path, "light tightness" (absence of extraneous light), mechanical rigidity to guard against dimensional changes and vibration of the optical system, and stability of the electronic components. The light transmission properties of the cuvettes is critical, especially in UV work where quartz glass cuvettes are essential. The worm drive mechanism that allows line (wavelength) selection must be free from backlash to permit settings to be repeated accurately and reliably. In scanning units in which the worm is driven by a motor, an $X-Y$ plotter should be used as the output device and not a linear strip-chart recorder. The X-axis (wavelength) voltage is obtained from a potentiometer driven by the worm. The Y-axis voltage is derived from the photodetector output. In this manner, an accurate $\%T$ or A curve versus λ is obtained. Some commercial units use a strip-chart recorder that is driven at constant speed by its own motor; the monochromator is driven by another motor. Usually the motors run at slightly different speeds and calibration is both difficult and unrepeatable.

10.2. INSTRUMENTATION FOR AVALANCHE PREDICTION MEASUREMENTS

Avalanche prediction is an area of growing importance not only because of safety requirements in recreational areas, but also environmental considerations associated with noise produced by jet aircraft. Avalanches may be triggered by a number of means, and sonic pressure or vibration is one such factor. As airlines seek to increase winter jet service into winter resort communities, avalanche study is becoming part of environmental impact statements. Some localities, such as Jackson Hole, Wyoming, have experienced increases in avalanche activity associated with sonic energy produced by aircraft. Avalanches are not only a hazard to developed areas, but to forested areas as well where considerable amounts of timber may be destroyed. This causes both direct economic loss and damage to water sheds.

A number of techniques are employed in avalanche prediction, not the least of which is observation of weather conditions before, during, and after snowstorms. In addition, two physical measurements are commonly carried out. These are the vertical temperature profile of the snowpack and the "hardness" of the snowpack in vertical profile. Normally the temperature measurement is conducted by digging a snowpit and inserting thermometers in the snow wall of the pit at various levels from top to bottom. This method is laborious, time consuming, and dangerous when carried out near an avalanche fracture zone. It is also subject to some error. From the measurements obtained, one determines whether the snowpack is isothermal, or displays a temperature gradient from the snow surface to the ground surface. Temperature gradients, particularly if large, are indicative of unstable conditions, while an isothermal pack indicates stability.

Snow "hardness" is usually measured using a mechanical device called a ram penetrometer. The device consists of an extendable shaft with a conical tip. The shaft is equipped with a mechanical stop, a concentric weight and a calibrated scale as shown in Fig. 10.8. The instrument is held loosely in a vertical position. The weight is raised to a measured height above the stop and released. The force of the weight striking the stop drives the ram into the snow. Frequently this process must be repeated several times before the ram moves. The process of dropping the weight and recording depth of penetration is repeated until the ram has penetrated to the ground surface. An empirical equation is used to calculate the snow "hardness" or ram resistance.

$$R_r = R + Q + nhR/\Delta$$

where R_r is the ram resistance, R is the weight of the movable ram weight (usually 1 kg), Q is the weight of the ram shaft (usually 0.5 kg/ section), h is the distance the weight falls (cm), n is the number of times the weight is dropped, and Δ is the penetration of the ram into snow (cm) for n drops of ram weight.

The results of the ram penetrometer test are plotted as shown in Fig. 10.9. What shows up are icy or crusty layers, and layers in which the snow has degenerated to form loose granular or depth hoar conditions. The hard icy layers provide surfaces on which the snow above may slide under appropriate conditions. The granular and depth hoar layers act as ball bearings and also produce sliding layers under appropriate conditions. Thus both icy (hard) layers and granular/depth hoar (soft "ball bearing") layers within the snowpack are potentially hazardous. There are many other factors to be considered in avalanche prediction. Those listed here are simply to provide a background for the designs of the two instruments subsequently presented in this section.

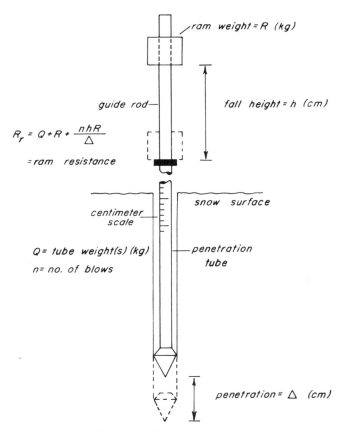

FIG. 10.8   Ram penetrometer.

There is an alternate ram design which uses a calibrated spring-loaded drive rather than a dropped weight. It is quicker to use than the conventional ram. Both the ram penetrometer and snowpit techniques require time, and the ram information is not immediately available. One must later draw the interpretive chart shown in Fig. 10.9. The time element places the personnel making these measurements in some jeopardy if they are working in a known avalanche region. The two instruments now to be described make the necessary temperature and relative R_r measurements directly, in short time, and with direct readout.

The temperature sensing probe, which eliminates the need for digging a snowpit is shown in cross-section in Fig. 10.10. Thermistor probes are spaced along the length of the main probe. These are alternately switched into a Wheatstone bridge resistance thermometer

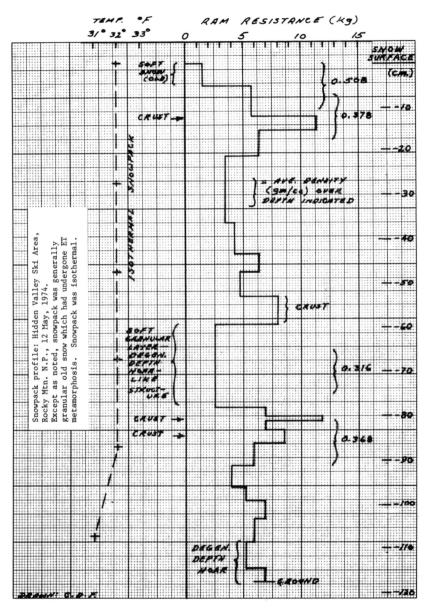

FIG. 10.9   Typical snowpack profile by ram penetrometer. Ram weight was 1 kg; tube weight was 0.5 kg. Ram resistance in kg was computed from the formula shown in the text.

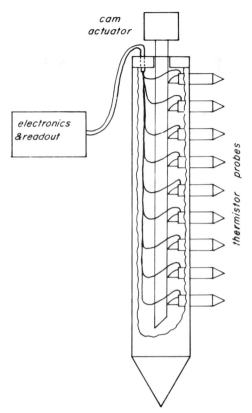

FIG. 10.10 Snowpack temperature probe. Partially cutaway view. The temperature probes are spring-loaded and retract into the body of the main probe unit when the cam is not actuated.

circuit. The main probe is inserted into the snow. A mechanical control at the top of the probe extends the thermistor probes into the snow surrounding the main probe. The operator then scans the thermistors with a switch and reads the temperatures from top to bottom of the probe. The thermistors are finally retracted into the main shaft and the probe is removed from the snow. Instead of a meter temperature read-out, an LED digital readout can be used. It is also possible to replace the manual switch by an automatic scanning (strobe) device. The factor here is weight versus convenience, since this is an instrument which must be backpacked into the field by a skier or someone on snowshoes.

The electronic ram penetrometer consists of a light aluminum tripod with pad feet to support it on the snow. The ram shaft is driven

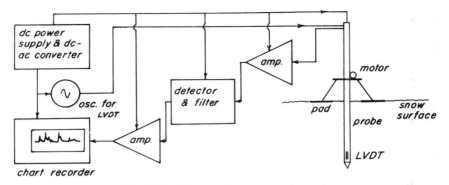

FIG. 10.11   Electronic ram penetrometer.

by a rack-and-pinion connected to a small motor. A cam and micro-switch attached to the motor gives an electrical impulse, which produces a vertical distance (depth) marker. The ram tip is connected to the spring-loaded core of a linear voltage differential transformer whose output is proportional to the resistance of the snow layer through which the tip is penetrating. The ram is driven at constant speed. The output of the LVDT is rectified, amplified, and displayed on a strip chart recorder. The depth markers are also displayed on the recorder by interrupting the LVDT trace momentarily. Thus one has an immediate graph of R_r versus distance. The LVDT spring transducer can be calibrated to give "absolute" values if they are required. Generally relative values of R_r suffice. The unit is battery operated and may be carried in a backpack. A thermal pen is used for the recorder. This system is shown in Fig. 10.11.

10.3.   MEASUREMENT OF STACK EMISSIONS

An important environmental problem is the monitoring of the stack emissions produced by stationary sources such as power plants, chemical plants, and cement plants. The major available technique, until recently, was the use of smoked glass filters. An observer compared filter density against smoke plume density. When the appropriate filter was found, it was referred to a numerical scale to find the associated Ringleman number. From the Ringleman number, one could make rough inferences concerning the amount of particulate matter dispersed in the plume. The Ringleman method was highly inaccurate and was

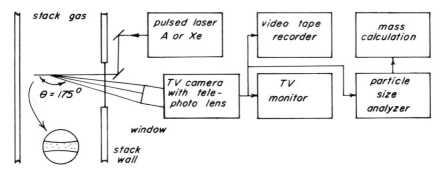

FIG. 10.12   Du Pont system for measuring stack-gas particle size and density. The circle in the figure indicates the volume defined by the intersections of the optical axes of the laser and the TV camera. This volume is approximately 1 mm³. System diagram redrawn with permission of E. I. du Pont de Nemours & Company, Inc.

subject to many errors, such as observer judgment, angle at which light struck the plume, sky background color behind the plume, and so on.

There are a number of laboratory instruments, including gas chromatographs and particulate size counters, which can be used for emission analysis, but they are not applicable to remote monitoring in the field. Recently, the Du Pont Company has announced a system which they developed to be able to comply with EPA standards in their plant operations. Their apparatus, which consists of a combined laser–television (LTV) monitor as shown schematically in Fig. 10.12, is a remote sensing system designed to monitor and analyze industrial particulates. In conjunction with a computer, it may be used for on-line continuous operation.

The unit monitors particulate size (0.01–10 μm diameter) and mass without regard to chemical composition. The theoretical basis for the instrument is Mie scattering theory, which relates apparent brightness of light reflected from a particle to size of the particle. A high-powered pulsed laser is used to illuminate particles flowing in stack gas. The pulse intervals are 0.5 μs. The focused laser light defines a sample volume on the order of 1 mm³. The light scattered from the particles in this volume is imaged in a long-focal-length TV camera tube. What appears on the TV monitor are distinct light flashes, each flash representing a particle. The electrical pulses produced by the camera tube (pulse per flash) are transmitted to a pulse analyzer.

The analyzer treats the data from each pulse of laser light as a camera frame and produces instantaneous data on the number and size

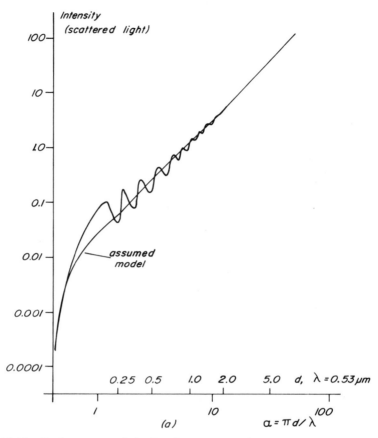

FIG. 10.13   Performance of the Du Pont system for particulate measurements (redrawn from *Du Pont Magazine*, November-December, 1976 College Supplement Insert). (a) Computer plot of scattered light intensity as a function of particle diameter; (b) Computer plot of scattered light intensity as a function of scattering angle with particle size as a parameter. Redrawn with permission of E. I. du Pont de Nemours & Company, Inc.

of particles in each frame. This information appears as a meter readout on a cathode ray tube, and gives a histogram of particle count versus size. The output can also be stored on magnetic tape for later analysis.

The sample volume is defined by the intersection of the TV camera tube's field of view and the path of the laser beam. Mass concentration at stack conditions is calculated and displayed from the measured particle size distribution and particle density. The scattered

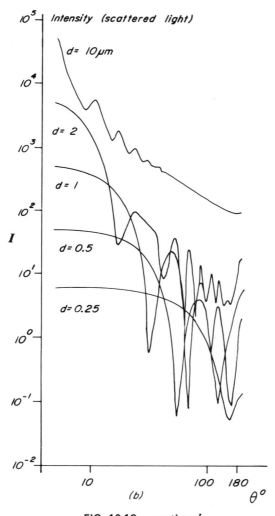

FIG. 10.13, *continued.*

light intensity varies with the angle between the camera tube and the laser beam, and the relative refractive index between the particle and its environment.

System performance is shown in Fig. 10.13, where $\alpha = \pi d / \lambda$, d is the particle diameter, I is the intensity of scattered light, θ is the scattering angle between incident light and detector, and λ is the wavelength of incident light.

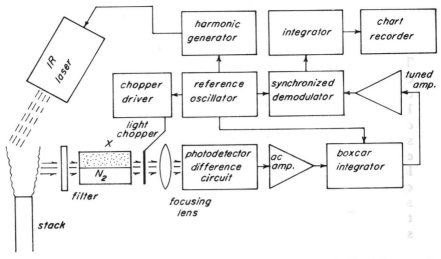

FIG. 10.14 NASA system for monitoring stack gases. One half of the sample chamber is filled with N_2 for reference calibration; the other half is filled with the suspected pollutant gas mixture indicated as X. Redrawn from *NASA Tech. Brief*, B75-10275, October, 1975.

Another system has been proposed by the Pasadena, California, office of NASA. It is shown schematically in Fig. 10.14. In this system, an IR laser is used to determine gaseous pollutants in stack plumes. The IR energy from the laser produces fluorescence in NO, NO_2, CO, CO_2, SO_2, and O_3. The detection system is designed to separate the fluorescence signal from background radiation, and to give a quantitative output representing the pollutant concentration. The filter is designed to absorb any radiation at the same wavelength as the laser energy. The two-compartment cell contains a pure sample of the pollutant to be measured in one chamber, and the other compartment contains a noninterfering gas such as helium or nitrogen. The pollutant-containing chamber absorbs any of the fluorescent energy radiated from the same gas in the plume. The difference in the intensity of the radiation leaving the two cell chambers is equal to the intensity of the fluorescent energy produced by the pollutant in the plume. When the radiant energy is converted in the photodetector to an electrical signal, the electrical signal is proportional to the concentration of pollutant in the plume.

The remainder of the electronic system shown in Fig. 10.14 contains signal processing circuits and synchronizing circuits for the pulsed output of the laser. The strip chart recorder displays pollutant concentration with time.

10.4. COMMUNICATION PROSTHESIS

The Bioengineering group at the University of Wyoming was contacted by the Gotsche Rehabilitation Center to develop a communication device for a 21-year-old male who had sustained permanent brain damage in an automobile accident. The clinical signs of this damage were as follows: loss of speech, moderate visual impairment such that characters under $\frac{1}{2}$ in. in height could not be read, damage to eye blink function, respiration normal but blowing and sucking ability lost, extreme loss of motor activity in limbs almost to the extent of quadriplegia. Hearing and other sensory abilities were apparently intact, as was mental acuity. The subject's impairment prevented the use of typical transducer–electric typewriter systems that use pneumatic switches or IR eye-blink transducers. The subject's manual ability was

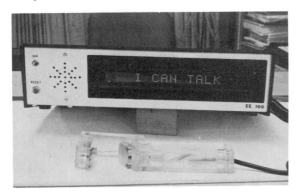

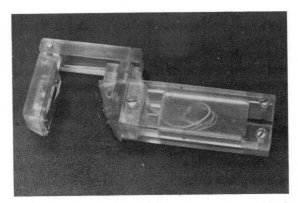

FIG. 10.15 Electronics unit and hand-held switching assembly for communication device.

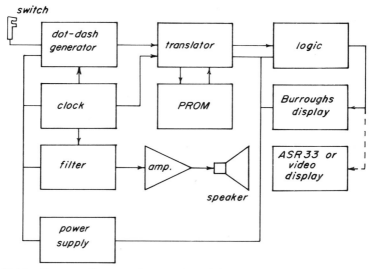

FIG. 10.16 System diagram for communication prosthesis. Arrows indicate information flow.

limited to motion of the left thumb and the ability to hold a lightweight object in his left hand.

While the subject was convalescing from the accident, he learned the Morse code and was equipped with a doorbell switch mounted on a block of wood. This in turn was connected to a battery and buzzer. With effort, he was able to communicate in slow Morse code. This, however, required that those around him understand the code.

The approach taken to this problem was the development of a hand-held switching device connected to an automatic dot–dash generator, similar to those used by radio operators. The device was designed so that downward thumb motion closed a contact activating one of the generators and upward motion closed a contact activating the other generator. With the thumb in rest position, neither contact was engaged. This arrangement eliminated the necessity for sending each dot and dash individually. The rate of dot–dash generation was made adjustable to suit the subject's requirements. An auditory signal was also generated so that the subject could hear his transmission.

To facilitate communication with others, a Morse-to-ASCII translator was designed using a PROM (programmable read only memory). The output of the translator was connected to a Burroughs Self-Scan display (as mentioned in Chapter 9) to give a visual readout. Connection may also be made to an ASR-33 teletype if hard copy is desired. This system is illustrated in Figs. 10.15 and 10.16.

10.5.   THERMOGRAPHY

An interesting application of temperature sensing transducers is their use in thermal scanning devices. Skin temperature is determined by remote scanning of the body surface or portions thereof. Applications include diagnosis of circulatory disorders and early detection of breast cancer in women. Tumors, because of vasculation, are hotter than surrounding normal tissue. The basis for a clinical scanning system is a sensitive IR detector. Human skin is an excellent radiator of IR energy.

Both slow- and fast-scanning systems have been developed. In the slow-scan system, a narrow aperture IR sensor is mounted in a scanning head. The body surface is scanned in a manner similar to the horizontal scan in a TV camera tube. The IR sensor output voltage is used to modulate the intensity of a light beam. The beam then scans a sheet of photographic film or projection paper. A line-by-line scanning regime is used to produce the complete thermogram of a portion of the body surface.

The fast-scan system utilizes a cathode-ray oscilloscope as the output device. The voltage from the IR sensor provides the horizontal scan. The voltage also intensity modulates the CRT electron beam (z-axis modulation). A vertical raster, as in television, of approximately 100 lines per frame is used to complete the full display. An advantage of this system, other than speed, is that signal enhancement can be used to bring out features in the gray scale of the display. Thus better resolution is available with this technique than with the slow-scan. The CRT display system is shown in block diagram form in Fig. 10.17.

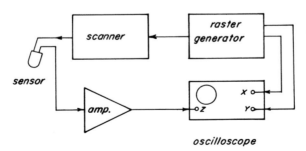

FIG. 10.17   Block diagram of a thermography system.

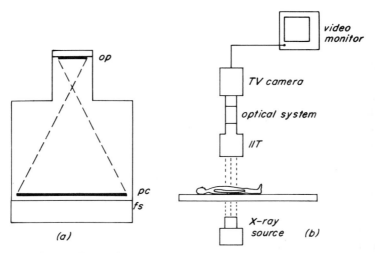

FIG. 10.18   Image intensification system: (a) schematic of image intensifier tube (IIT): fs = fluorescent screen; op = output screen (image); pc = photo-cathode; (b) overall system.

10.6.   X-RAY INTENSIFICATION AND ISOTOPE SCANNERS

An example of the application of modern instrumentation to a long-standing clinical-diagnostic technique is X-ray image intensification. The hazards associated with normal X-radiography are significantly reduced since the radiation levels are reduced. The basic system is illustrated in Fig. 10.18. In certain respects the image intensification tube (IIT) is similar to a photomultiplier tube. The X-ray image is focused on a fluorescent screen within the IIT. The image is sensed by a photocathode matrix, amplified, and projected on an output screen integral with the IIT. The output image is picked up by a TV camera and displayed on a closed-circuit video monitor. Alternatively, a cine camera may be used in place of the TV camera if a permanent record is required. The advantages of such a system are low radiation level, immediate display (no X-ray plates to be developed), and safety to operating personnel because of the closed-circuit TV operation. The main disadvantage of such systems is size and lack of portability, although systems of smaller physical size are being developed.

Isotope scanners are based upon the principle that certain organs, such as the thyroid, take up particular substances. If these substances

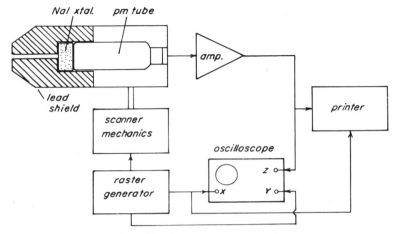

FIG. 10.19   Isotope scanner system.

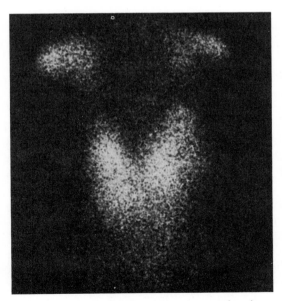

FIG. 10.20   X-ray film output from an isotope scanner showing a normal human thyroid. The figure is the reverse image from the original. Scan courtesy of K. C. Rock, University of Colorado Medical Center, Denver, Colorado.

are tagged with a radioisotope, then organ function can be monitored by the amount of gamma radiation emitted by the organ. The organ region is remotely scanned in much the same manner as in thermography, as discussed in a prior section of this chapter. The scanning head, or goniometer, consists of a lead shield equipped with a small aperture to pass γ radiation. Behind the aperture is a NaI crystal backed by a photomultiplier tube; this arrangement is a scintillation counter. The γ radiation, when absorbed in the NaI crystal, emits light that is registered by the PM tube. The light produced, and hence the electrical output of the PM tube, is proportional to the energy of the γ radiation. The output from the PM tube is amplified and directed to a pulse height analyzer, as shown in Fig. 10.19. The output from the pulse height analyzer may be used to intensity modulate the electron beam in a cathode ray oscilloscope. In this application, a vertical raster determines the number of scanning lines and the intensity-modulated horizontal display provides information.

In an alternative arrangement, the output of the pulse height analyzer is directed to a bar printer. A horizontal scan produces bars, which in number, are proportional to the intensity of the γ energy. A vertical raster generator drives both the scanner and the printer to produce a complete output, as shown by the thyroid scan in Fig. 10.20.

10.7.   MULTISPECTRAL IMAGING

Photographic techniques using infrared, visible, and ultraviolet light have for some time been employed in clinical studies to diagnose and follow the courses of various diseases. Recently V. J. Anselmo of Caltech/JPL, Pasadena, California, with NASA support, has developed a multispectral imaging technique for diagnostic analysis of skin damage in burn cases. The basic theory behind this system is differential absorption by and penetration into skin layers as a function of light wavelength. Vasculation and physiologic condition of the dermal layers can be estimated to assist in diagnosing burn depth, tissue damage, and rate of healing.

The overall system is shown in Fig. 10.21. In operation, the subject is illuminated by broad spectrum lamps. The scanning mirror directs the light reflected from the patient through a lens system to three imaging detectors. Each detector is equipped with a filter such that data are obtained for three spectral ranges of interest. Each detector is a linear array of silicon photodiodes and produces image data as the mirror scans the patient. The three reflected light images are

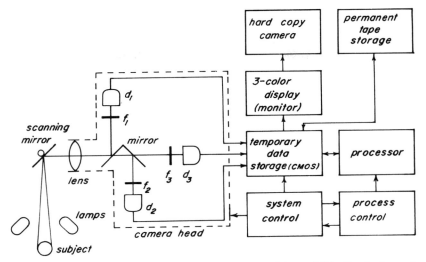

FIG. 10.21   Multispectral imaging system developed by NASA. Dashed portion of diagram indicates components contained within the camera head; $f_1 - f_3$ = filters; $d_1 - d_3$ = photodetectors. Redrawn from *NASA Tech. Briefs*, Winter, 1976; work conducted by V. J. Anselmo of Caltech/JPL.

digitalized and stored. This function is carried out in a processor which, in turn, is controlled by a microcomputer.

Brightness histograms are calculated for each image and these data used to select contrast enhancements. As the final step, the processor computes spectral image ratios, which are then displayed as the false color images used for diagnosis. Output is in three forms: 3-color CRT display, film (camera), and digital data on tape.

For burn analysis, full-thickness injury appears white and yellow in the display, deep partial-thickness injury is red, and shallow partial-thickness injury is blue.

10.8. TEMPERATURE SENSOR

Some complete instrument systems are rather simple such as the temperature sensor, shown in Fig. 10.22, developed through NASA Langley Research Center. The temperature sensor is a Fairchild FDH600 silicon diode in which the forward voltage across the diode is nearly a linear function of temperature from 77 to > 300 K, provided that forward current in the device remains constant. At a bias current of

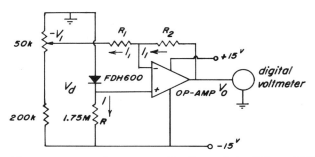

FIG. 10.22   Temperature sensing system. R_1, R_2, and the 50 k and 200 k resistors should be temperature compensated (Low Tempco). R_1 and R_2 are selected to provide a suitable voltage value V_0 for the digital voltmeter. Redrawn from *NASA Tech. Brief*, B75-10260, October, 1975.

8.5 μA, the diode is linear to $\pm 0.1°C$ over the range 0–40°C (273–313 K). Pertinent circuit analysis is (symbols from diagram):

$$I \sim (15 - V_d)/R \sim 8.5 \,\mu A$$

$$I_1 = \frac{V_d + V_1}{R_1} = \frac{V_0 - V_d}{R_2}$$

thus

$$V_0 = V_d(1 + R_2/R_1) + (R_2/R_1)V_1$$

System tests reported by NASA indicate that a 1% change in diode current produces a 0.4 mV change in output voltage. An additional operational amplifier can be added to the circuit to provide constant current regulation of the diode current.

10.9.   CONCLUSION

In this chapter, we have presented examples of typical instrumentation systems designed to meet specific clinical and environmental measurement requirements. The examples were selected to indicate the broad spectrum of designs and techniques found in the area encompassed by this book.

10.10.   REFERENCES

Ackermann, P. G., 1972, *Electronic Instrumentation in the Clinical Laboratory*, Little, Brown, Boston.

Cromwell, L., F. Weibell, E. Pfeiffer, and L. B. Usselman, 1973, *Biomedical Instrumentation and Measurements*, Prentice-Hall, Englewood Cliffs, N.J.

Jacobson, B., and J. G. Webster, 1977, *Medicine and Clinical Engineering*, Prentice-Hall, Englewood Cliffs, N. J.

Thomas, H. E., 1974, *Handbook of Biomedical Instrumentation and Measurement*, Reston Publishing Co., Reston, Virginia.

Wolff, H. S., 1970, *Biomedical Engineering*, World University Library, McGraw-Hill, New York.

Du Pont Magazine, November-December, 1976.

NASA *Tech. Briefs*, 1975–1977.

A useful source reference for environmental monitoring instruments is the following: *Survey of Instrumentation for Environmental Monitoring* in four volumes (Air, Water, Radiation, Biomedical), available from the Technical Information Department, Lawrence Berkeley Laboratory, University of California, Berkeley, CA 94720. This is not a text book, but a compendium of technical specifications for commercially available monitoring equipment. Certain tests are also described and defined.

Section 5

Telemetry

11

Telemetry Systems

11.1. INTRODUCTION

In the late 1950s, techniques began to emerge for transmitting physiological information from within the human body by way of radio links.
The devices were called endoradiosondes, a term adapted from the
atmospheric balloon radiosondes used for many years to transmit
data utilized in weather prediction. These efforts were the beginning
of the field of biological telemetry.

Early attempts were directed toward obtaining diagnostic information that could not be otherwise obtained. Although this is still
an important aspect of the field, it has branched out into a number of
other areas, to the extent that the medical affiliation is somewhat
obscured. Much of the current work is directed toward monitoring
wildlife behavior and physiology, particularly with respect to the
affects of human encroachment upon the environment. Applications
also relate to physiologic studies in domestic livestock.

There are some rather different problems associated with the use of
telemetry in human subjects as opposed to use with animals, especially
free-roaming wild animals. Normally human subjects are in a restricted

geographic area defined by a clinical facility in a hospital or medical center. Wild animals, on the other hand, may be in a penned pasture or confine, but more usually are free roaming over their natural range, which may entail hundreds of square kilometers. Thus there are some rather different design considerations for clinical laboratory telemetry equipment versus equipment used in wildlife studies.

Two general classes of telemetry packages are used: those which are external to the subject and are held in place by some sort of strap or harness, and those which are implanted within body tissues. External units do not require surgical procedures and may be serviced easily; however, they are subject to mechanical damage and various electrical and mechanical artifact phenomena. Implantable units require surgery on the subject and are difficult to service, but operate in a nearly constant environment essentially immune to mechanical trauma and artifact.

In this chapter, we will examine some of the systems and techniques used to obtain physiologic data from remote subjects. For the most part, general design considerations will be examined rather than detailed presentation of circuits. With the very rapid development of microcircuit technology, many designs are now possible that would be either impossible or impractical with discrete component assemblies.

11.2. THE BASIC TELEMETRY SYSTEM

As illustrated in Fig. 11.1, the basic radiotelemetry system consists of the following functional components: data acquisition module, modulator, radio transmitter, transmitting antenna, transmission medium, receiving antenna, telemetry receiver, and data processor. In

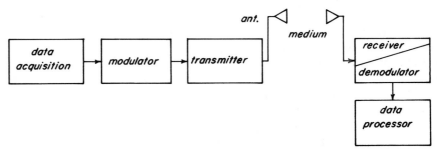

FIG. 11.1   Basic telemetry system.

any given system, not all of these functional modules are necessarily present. Many of the early wildlife telemetry systems were designed simply for tracking the position of an animal as it ranged through its habitat. The only information obtained was the physical position of the animal. In this case, only a very simple system, consisting of a radio transmitter that emitted either continuous or pulsed radio energy with a fixed center (carrier) frequency, was required. A simple radio receiver tuned to the transmitter frequency, and a transmitting and a receiving antenna completed the system. Air was normally the transmission medium. The first refinement of this system was the use of a polarized transmitting antenna and a direction-finding type of receiving antenna to allow accurate triangulation of the animal's position. The strength of the received signal gave an indication of the distance of the animal from the receiving (tracking) antenna. Tracking was accomplished either on foot or horseback, or by air.

Telemetry of physiological information requires more system sophistication than simple tracking. Consider the ECG telemetry system illustrated in Fig. 11.2. This system is used in exercise physiology studies to monitor the electrocardiogram waveform. Data acquisition consists of electrodes (two or three, minimum) placed on the subject and an amplifier–filter module to provide a "clean" signal of sufficient amplitude to drive the modulator. The demodulator in the receiving system provides a signal for immediate display on an oscilloscope (or chart recorder), and a signal for recording on magnetic tape for subsequent computer analysis. In some systems, on-line computer analysis is carried out and magnetic tape storage is not used.

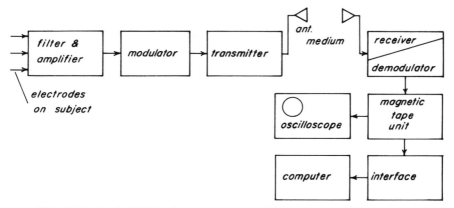

FIG. 11.2   Basic ECG telemetry system with receiver signal processing.

The two examples discussed thus far are single-channel systems. Transmission of several physiological signals over a single channel, or tracking of several animals using the same carrier frequency requires some sort of signal multiplexing. This subject will be taken up in subsequent sections of this chapter.

11.3. BASIC DESIGN CONSIDERATIONS

There are several fundamental factors that must be examined in the design of any telemetry system. These include, but are not limited to: range, power consumption, physical size, weight, radio frequency of operation, duration of use (lifetime), environment in which device is to operate, transmission medium, and subject animal.

11.3.1. Range

The distance over which a telemetry system must be capable of operating depends strongly upon application. Generally, systems used for either clinical diagnostic or research studies need to operate only over distances of a few meters at most. Systems used to monitor athletes during exercise may need an effective range of several hundred meters. Units used in monitoring free-roaming animals often must operate over a minimum distance of several kilometers.

11.3.2. Power Consumption

The power consumed by a telemetry system, for fixed carrier frequency, is normally proportional to the distance of the transmission path and the number of data channels. Power drain from the power source is then proportional to the amount of power radiated from the transmitting antenna. Generally, the longer the range of transmission, the greater the power required. Other factors to be considered are carrier frequency and antenna gain (directivity), both of which affect transmission efficiency.

Telemetry transmitters are usually battery operated, and frequently the receivers as well. It is normally not necessary to have a telemetry system operating on a continuous basis for 24 hours, and in some cases, the system may not be used every day. If the system can be turned off when not in use, battery life (and hence system life) can be extended substantially. Incorporated in many systems are radio operated switches that allow the telemetry unit to be switched on or off by a remote-controlled radio pulse. With tracking transmitters used on wildlife, it is sometimes desirable to have continuous system operation. To preserve battery life, pulsed rather than continuous wave (CW) transmission is used. A typical duty cycle (on time vs off time) is 1–3%.

Further power saving can be achieved by careful circuit design to insure minimum current drain in each of the system components. Hybrid microcircuit design usually results in much lower current drain than occurs with discrete component design. Batteries are normally rated in terms of ampere–hour capacity at room temperature, derated as temperature decreases. Thus one must design circuitry keeping in mind the temperature range over which the system is to operate.

Mercury or lithium batteries are normally used in telemeters that are implanted in animals, since essentially a constant temperature environment exists. Mercury cells are quite temperature sensitive and the current output drops to nearly zero just below 0°C. Animal body temperatures are typically on the order of 37°C. On the other hand, mercury cells are small and produce a nearly constant output voltage over their lifetime. Lithium batteries are gradually replacing mercury cells, but they are not yet available in very small physical packages. For telemetry packages that are exposed to extremely low temperatures, either lithium or alkaline batteries are used.

The transmission medium also affects power consumption. Air is the usual medium, but a portion of the transmission channel may be water (telemetry from fish or other aquatic animals) or animal tissue in the case of implants and endoradiosondes. This subject will be treated in more detail in a subsequent section.

11.3.3. Physical Size

This subject needs little elaboration. Telemetry equipment that is to be implanted in an animal generally needs to be as small as possible, regardless of animal size, to insure minimum intrusion upon the living system. Radiosonde "pills" must be swallowable. Equipment that is not implanted or swallowed may be of larger size, but it should not be so large or heavy that it interferes with the normal activity of the subject. Usually size requirements can be met by hybrid microcircuit design. Much of the volume of modern telemetry transmitter packages is related to the batteries rather than the electronic circuitry.

11.3.4. Weight

Animal or subject size generally dictates the maximum weight of a telemetry transmitter package. With modern electronics, batteries are by far the major contributor to weight. When one is working with fish, such as trout, or birds or small mammals, the telemetry packages must be small and light. The designer will generally have to compromise battery and telemeter lifetime in order to produce a telemetry package that does not interfere with the normal motion and activity of the subject. The shape of the telemetry package is also important. It must be designed so that it does not interfere with the subject's physiology and activity, and it should not be vulnerable to mechanical damage.

11.3.5.   Radio Frequency of Operation

Telemetry equipment must be operated in accordance with the regulations of the Federal Communications Commission. Certain frequency bands have been designated by the FCC for scientific and medical use, and for use by government agencies (state and federal). Generally wildlife telemetry falls under the last category since state game and fish departments, the U.S. Forest Service, or other government agencies are involved.

Other frequencies may be used provided that the signals radiated do not interefere with authorized users at the same frequency. This is generally interpreted as radiation at power levels of 100 mW or less; however, there is no definite power level specified. In designing telemetry systems, one should avoid all navigational aid frequencies, designated emergency frequencies, and any frequencies on which interference by the telemeter might endanger public welfare or safety.

In most cases, one selects frequency bands of low utilization to avoid spurious signals from interfering with the telemetry system. Normally one should contact the FCC for information since the rules and regulations change periodically.

Selection of frequency for maximum transmission efficiency will be treated subsequently.

11.3.6.   Lifetime

Two factors enter into the determination of the lifetime of a telemetry package. One factor is current drain on the batteries and battery maximum size and weight limitations. The other factor is the time period over which telemetry monitoring is required. Clinical monitoring of a human subject normally lasts from a few minutes to a few days. Wildlife tracking, on the other hand, usually requires information for up to a year or more. Thus there must be some design tradeoff between package size and weight, its lifetime, and the nature of the data obtained.

11.3.7.   Environment

Telemetry transmitters are required to operate under a number of environmental conditions. Many are implanted in animals where they are subject to attack by body fluids, although temperature remains essentially constant. Some may operate in either fresh or salt water attached to aquatic animals. The corrosive nature of a wet environment and possible mechanical damage must be considered in the package design. Wide variations in temperature may also be encountered. Swallowable radio "pills" must be protected against the corrosive action of digestive fluids. External tracking beacons used on large wild animals such as bear, deer, and elk must operate under rather harsh environmental conditions. They are subject to temperature extremes, rain and snow, and mechanical trauma. They must be designed to withstand considerable shock and vibration and should not be vulnerable to intentional or accidental removal by the animal. Usually various neck collar designs satisfy these requirements.

Most telemetry packages must be encapsulated in some protective material to seal them hermetically against adverse environments. Silastic rubbers, fiberglass, epoxys, and hydrophobic waxes are used. Frequently layers of differing materials are used. There are two main requirements: 1. The materials used must resist moisture penetration and corrosion. 2. The materials must be nontoxic and compatible with body tissues when the units are implanted.

11.3.8.   Transmission Medium

The normal transmission medium for telemetry systems is air. This presents no particular radio-wave propagation problems. When telemeters are used on aquatic animals or are implanted within an animal, then some substantial propagation problems develop. The media themselves are electrically conductive and there is considerable power loss during transmission through the media.

A second problem is the air–medium interface, either air–water or air–tissue, and in some cases, air–water–tissue. Since radio waves are transverse electromagnetic (TEM) waves, they are subject to various reflection and refraction phenomena. TEM conditions hold only beginning several wavelengths from the antenna (far field). In the medium immediately surrounding the antenna, the electromagnetic field distribution is extremely complex and not generally amenable to mathematical analysis. All of these factors reduce the amount of power available to the receiving antenna relative to the input power supplied to the transmitting antenna. Polarization of the transmitting antenna relative to polarization of the receiving antenna also affects the channel transmission efficiency. These matters are treated in some detail in a subsequent section of this chapter. Generally speaking, increased transmission power, with concomitant battery drain, is required to overcome the effects of non-air transmission media.

11.3.9. Subject Animal

The signal source, human or animal, has been alluded to several times above. In certain aspects, the design of systems to be used with human subjects is less critical than the design of animal telemeters. Normally humans can be instructed in the "do's" and "don'ts" of using and handling of a telemetry package. Systems for animal use must be designed so that the animal cannot damage them either by direct action or abrasion against trees, rocks, or other physical objects. Size, weight, and environmental considerations have been discussed previously.

11.4. VARIABLES TO BE MONITORED

There are several "standard" variables that are monitored by telemetry systems including heart rate, ECG, EEG, respiration rate, gilling rate of fish, temperature, pressure, activity, and position or location. With appropriate data-handling systems, information can be obtained to differentiate between waking and sleeping states of animals. Systems have also been developed for measuring pH, blood oxygenation, and other physiologic variables.

The transducers used to acquire the desired signals may include metal electrodes, electrochemical electrodes, pressure transducers, thermistors, and strain gauges. Tracking beacons require no transducers. Typically a blocking oscillator or similar circuit is used to produce a pulsed rf output.

11.5. BASIC CIRCUITS—HISTORICAL DISCUSSION

Figure 11.3 illustrates a basic circuit for a beacon transmitter used in wildlife tracking (Lonsdale, 1969). The circuit is a modified Hartley oscillator that operates in the self-blocking mode. The double parallel

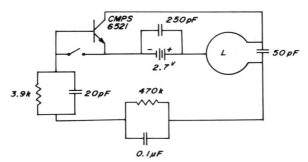

FIG. 11.3 Lonsdale beacon transmitter (fish "tag"). $L = 1$ turn on 2.86 cm diameter form, center-tapped; no. 14 or 16 AWG wire may be used. Switch is open for normal operation.

RC circuits set the pulse repetition rate and the pulse duration rate. The switch is a magnetic reed switch used to activate the unit when it is attached to the animal. This particular unit was designed for free-swimming trout so that the oscillator coil L also serves as the antenna. The unit operates in the 100 mHz band with a pulse output power of 0.75 mW. The oscillator coil is a single turn of wire 2.86 cm in diameter or 17.97 cm in length. Since transmission is from water (dielectric constant ~ 80) to air, the effective length of the antenna is $1/\sqrt{80}$ times the air length. Thus the effective air length is nearly 162 cm or slightly more than one-half wavelength at 100 mHz. This is a rough calculation since the antenna is a center-tapped loop rather than a linear dipole.

The pulse rate is 50 Hz with a 10% duty cycle. For these parameters, the unit has a range of 0.3 km for transmission from beneath the water surface (fresh water) to an above-surface receiver. The batteries in the original design were four 1.35 V, 105 mAh silver oxide cells connected in series–parallel. The beacon has a lifetime of approximately four weeks.

A commercial portable FM receiver is used in conjunction with a two-element yagi antenna. The antenna must be positioned with the dipoles oriented vertically.

Figure 11.4 shows the output of this telemeter operating in air. The oscilloscope tracings are obtained from the audio amplifier output of a commercial FM receiver. For operation in air, the pulse repetition frequency is 111 Hz and the pulse duration including the "ringing" is 4 ms. The true pulse is much distorted by the FM receiver. Component values as shown in Fig. 11.3 were used.

There is some oscillator frequency drift with this system, but the tuning of the FM receiver is broad enough to allow tracking of the

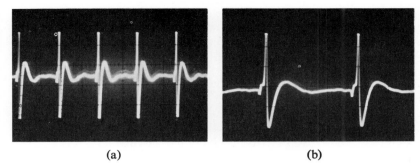

(a)   (b)

FIG. 11.4   FM receiver output signal for Lonsdale beacon transmitter. (a) General trace. (b) Expanded trace to show signal distortion introduced by receiver. Pulse rate = 111 Hz; pulse duration including "ringing" = 4 ms.

signal. Figure 11.5 shows a crystal controlled transmitter for improved frequency stability (Slagle, 1965).

The unit shown in Fig. 11.3 is encapsulated with several layers of fiberglass which, in turn, are coated with latex or silastic rubber. The package is in the form of a saddle that fits dorsally on the fish anterior to the dorsal fin. A metal pin passed through the flesh just above the spine holds the unit in place. This same unit could be used, with a separate antenna, on terrestrial animals. The telemeter weighs 20 g.

An early circuit that permitted transmission of useful data other than location is the pressure-monitoring radiosonde of MacKay, 1960, shown as Fig. 11.6. The pressure transducer is also the oscillator coil. The coil is wound on a cylindrical lucite form, closed on one end. The

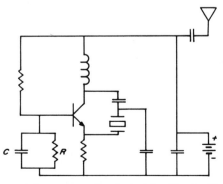

FIG. 11.5   CW beacon telemeter as developed by Slagle. For pulsed operation, R is omitted and C is selected for the desired pulse rate.

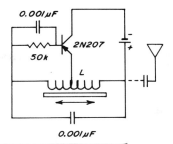

FIG. 11.6   MacKay pressure endoradiosonde. Antenna is optional; normally, direct radiation from the coil (*L*) would be used. *L* = 250 turns of AWG no. 40 wire. Coil is center-tapped.

coil form is concentric with a small ferrite rod attached by one end to a thin rubber diaphragm and free on the opposite end. The diaphragm is cemented to the open end of the lucite cylinder. This forms a closed pressure transducer with the ferrite as a piston. Changes in external pressure cause the diaphragm to move, which displaces the ferrite relative to the coil. This in turn changes the frequency of oscillation of the circuit. The telemeter output is thus a frequency shift in proportion to pressure. This is a CW unit rather than a pulsed-output device.

If the diaphragm is assumed to be linear, then the change in inductance ΔL is directly proportional to the position change of the ferrite Δl, which in turn is directly proportional to the pressure change Δp. Assuming no stray inductance or capacitance effects in the oscillator circuit, then the frequency shift is given by:

$$f_0 = 1/(2\pi\sqrt{LC}) = \text{quiescent frequency}$$

$$f_1 = 1/[2\pi\sqrt{C(L + \Delta L)}] = \text{new frequency for change in inductance } \Delta L$$

$$\Delta f = f_0 - f_1 = \frac{1}{2\pi\sqrt{LC}}\left[1 - \frac{1}{\sqrt{1 + (\Delta L/L)}}\right]$$

if $\Delta L \ll L$, then we can use the binomial series expansion to find:

$$\Delta f \sim \frac{1}{2\pi\sqrt{LC}}(\Delta L/2L) = f_0\,\Delta L/2L$$

or if

$$\Delta L = k\,\Delta p$$
$$\Delta f \sim f_0 k\,\Delta p/2L$$

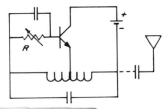

FIG. 11.7 Simple temperature telemeter. R is a thermistor, and when R varies with temperature, the squegging rate of the modified Hartley oscillator changes accordingly.

or

$$\Delta f \sim K \, \Delta p$$

where

$$K = f_0 k / 2L$$

A telemeter to transmit temperature variation is shown in Fig. 11.7. A thermistor is used as the temperature transducer. The circuit is a modified Hartley oscillator operating in the blocking mode. The pulse repetition frequency (prf) is proportional to temperature. Pressure data could also be transmitted by this technique using a pressure-sensitive resistive element.

An early two-channel telemetry system developed by MacKay, 1959, is shown in Fig. 11.8. The variable-inductance pressure transducer is the same as that described above. The circuit is modified for blocking oscillation by insertion of a thermistor. Temperature is determined by the squegging rate and pressure by the shift in the center frequency of the oscillator.

To receive the signals from the MacKay swallowable endoradiosondes, a flat multiturn loop antenna was used (100 turns, 5 cm in

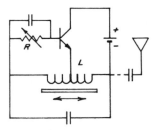

FIG. 11.8 MacKay dual-channel telemeter: R varies squegging rate with temperature (thermistor); L varies base frequency of oscillation with pressure.

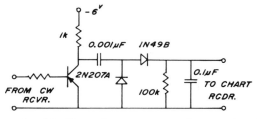

FIG. 11.9   Detector for MacKay telemeters to produce analog signal for a chart recorder.

diameter). A CW receiver operating over the range 190–550 kHz served as the detector. Audio output from the BFO in the receiver could be used to determine pressure or temperature variations. For a permanent record, the circuit shown in Fig. 11.9 served as a demodulator.

A more sophisticated temperature telemeter circuit has been issued by the Pasadena Office of NASA, and is shown in Fig. 11.10. In this case a modified Colpitts oscillator is used that incorporates thin-film and chip components. The thermistor temperature sensor is composed of a nickel thin-film deposited on a nickel substrate. This fabrication yields rapid response and great sensitivity. The system is basically FM. The change in thermistor resistance with temperature causes a change in the reverse voltage supplied to the varactor diode, which in turn changes the effective capacitance in the oscillator frequency-determining circuit. The potential at point A must remain negative relative to point B to maintain reverse bias on the diode. The diode capacitance varies with reverse voltage according to

$$C = \text{constant } V_r^{-\alpha}$$

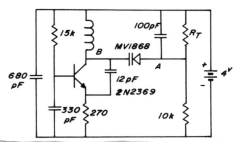

FIG. 11.10   Temperature telemetry transmitter developed by NASA. An MV1868 varactor is used (between points A and B on the schematic). Redrawn from *NASA Tech. Brief* NPO-10649.

where α is a property of a particular diode type. The voltage at point A is given by

$$V_A = \frac{10k}{R_T + 10k} \quad (4)$$

$$V_r = 4 - V_A = 4\left(\frac{R_T}{R_T + 10k}\right)$$

Thus

$$C = \text{constant} \left(\frac{4R_T}{R_T + 10k}\right)^{-\alpha}$$

This means that the oscillator frequency is generally not linearly proportional to changes in temperature since the frequency is given by

$$f = 1/(2\pi\sqrt{LC})$$

The NASA circuit is designed to operate at a nominal 115 MHz. The

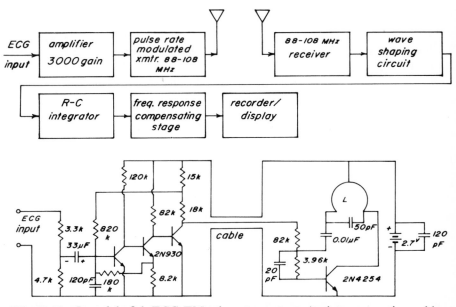

FIG. 11.11   Lonsdale fish ECG FM telemetry system. An interconnecting cable is used between the amplifier and transmitter; $L = 1\frac{1}{2}$ turn on $\frac{7}{8}''$ diameter form, center-tapped.

sensitivity is given by

$$S = \Delta f/\Delta T \sim 25 \text{ kHz/}°F$$

A somewhat more sophisticated circuit for telemetering the ECG signal from free-swimming fish was described by Lonsdale, 1969, and is shown in functional and schematic form as Fig. 11.11. The 3000 gain in the input amplifier was required to bring the 200 μV fish skin ECG potential up to a suitable level to drive the transmitter. Pulse rate modulation was used (pulse rate proportional to voltage amplitude of ECG). Two stainless steel pickup leads were placed on the skin of the fish just over the heart. To eliminate the effect of contact potentials, 33 μF coupling capacitors were used in series with each lead.

The receiver was a commercial portable FM unit. The detector output was fed into a wave-shaping circuit (Schmitt trigger) to produce sharp pulses of uniform profile. The trigger in turn was used to drive

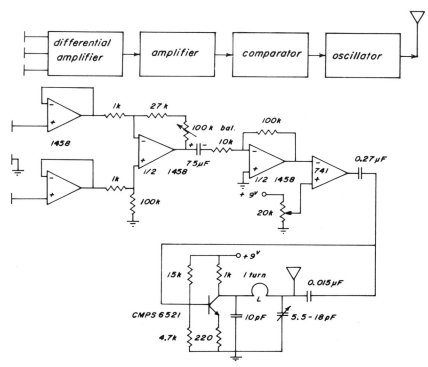

FIG. 11.12   Transmitter for human cardiotachometer: L = 1 turn of AWG no. 14 or 16 wire, approximately 1″ in diameter. Diameter depends upon desired operating frequency of oscillator. The op-amps are operated from ±9 V; bias connections are not shown.

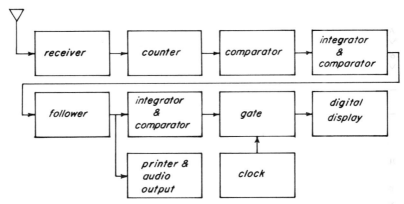

FIG. 11.13   Block diagram of receiver and digital system for human cardio-tachometer system.

an emitter-coupled monostable multivibrator to generate standard-width, standard-height 50% duty cycle pulses. Signal detection occurs in the two-stage *RC* integrator. An *RC* peaking circuit is then used to restore low frequency components lost in the original modulation process.

A cardiotachometer system, developed by Ferris and Hakes, 1975, is shown in Fig. 11.12. This system was designed to monitor the heart rate of exercising human subjects. Operational amplifiers are used where possible. Follower circuits are used to match the electrodes to an input differential amplifier which, in turn, drives a voltage comparator. The comparator converts the analog R wave into a rectangular pulse. This pulse in turn triggers an oscillator tuned for the range 70–90 MHz. The oscillator drives a dipole antenna. An AM receiver and digital system, as shown in Fig. 11.13 completes the system.

The systems described so far have used discrete components and some integrated circuit chips. A more modern approach will be described in the following section.

11.6.   CURRENT DESIGN TRENDS

The approach to modern telemetry design, in the biological sciences, is the use of microcircuit and hybrid microcircuit technology. Using these techniques, it is possible to reduce substantially both physical size and power demand in telemetry transmitters.

There is really very little improvement required in beacon telemetry packages, other than to reduce size and improve efficiency. One hybrid

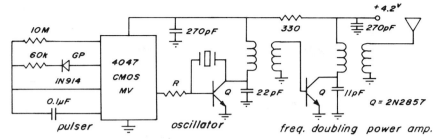

FIG. 11.14 Hybrid microcircuit telemeter: R is adjusted for desired power output of telemeter; a typical value is 68 k.

microcircuit design has been described by Weeks *et al.*, 1976. It is incorporated into a tranquilizing dart used to immobilize a large animal. Typically the animal, such as an elk, may be darted from an aircraft. The problem that arises is to find the downed animal after the aircraft has landed, or to find the dart if the shot was a miss. Either animal or dart may be obscured in heavy brush and not easily located by visual methods.

The telemeter is shown in Fig. 11.14. The following components are included: chip transistors, diodes, monolithic capacitors, monolithic digital integrated circuit, thin film resistors (tantalum nitride), miniature hand-wound toroids, and crystal. Alumina substrates were used with the components bonded by thermocompression using gold wire or by use of silver-impregnated, conducting epoxy resin. The circuit inter-connection pattern was produced by standard thin film negative photoresist techniques applied to gold-film-coated alumina substrates. The overall package measures 1.65 cm long by 0.76 cm high by 0.76 cm wide, and is coated with "Humi-seal" to provide a moisture proof coating. The specifications for the device are:

Average current drain	0.1 mA
Expected battery life	
(Type RM 212 mercury cells)	150 h
PRF	1 Hz
Pulse width	10 ms
Weight	2 g
Dimensions	1.65 × 0.76 × 0.76 cm
Operating temperature range	−4 to +52°C
Operating frequency	164.5125 MHz

Several options exist for the antenna. Although rather limited range results, one may simply use the radiation from the coils. Other-wise the metal casing of the dart can be used, or a wire antenna attached along the shaft of the dart.

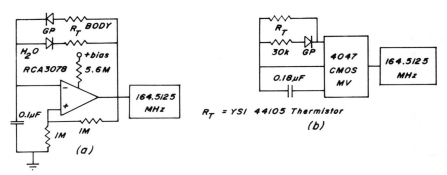

FIG. 11.15   Systems using telemeter of Fig. 11.14 to transmit body temperature data from aquatic animals. The circuit shown in (a) yields data on fish body temperature and water temperature. The circuit in (b) yields a single channel of temperature data. The block designated 164.5125 MHz is the transmitter of Fig. 11.14.

An interesting application of this transmitter is in the system described by McAleenan *et al.*, 1976, for monitoring temperature in free-swimming fish. Two designs are shown in Fig. 11.15. That shown in Fig. 11.15(a) permits monitoring of the water temperature and surface temperature of the fish. The unit is packaged in a manner similar to the fish-telemeter saddle pack described in the previous section. The water-temperature sensing thermistor is on the outside of the package and the fish temperature sensor rests against the fish on the inside of the package. The thermistors are Yellow Springs Instruments type 44015 and an RCA 3078 low-current operational amplifier is used. The pulser circuit is a conventional asymmetric multivibrator. The duration of one-half of the output is controlled by the water temperature thermistor and the fish's skin temperature controls the duration of the other half cycle.

Figure 11.15(b) is an implantable version for monitoring internal temperature of a free-swimming fish. The 4047 is an astable multivibrator chip. The longer portion of the period is controlled by the thermistor and the remaining portion is fixed by the resistor used for pulsing the radio transmitter.

The package was coated with "Humi-seal" overlayered with medical-grade silastic rubber. No antennas were used in either case, as the monitoring was done in a laboratory environment with the fish in large aquaria.

This same basic transmitter design (Fig. 11.14) has been used successfully in aerial tracking of the Lahontan Cut-throat Trout in the

Truckee River near Reno, Nevada by Weeks *et al.*, 1977. A half-wave folded-dipole antenna was constructed as a strip around the outer edge of the fish telemetry package. For the 164 mHz US Fish and Wildlife Service telemetry band used, the length of the antenna in water was approximately 10 cm (90 cm in air). Telemetry antennas will be discussed in section 11.8.

Although little has changed in the basic design of tracking beacons, there have been some considerable advances in the signal processing of the received signal. When a number of animals are being tracked in a given geographic area, it is not always possible to assign a separate carrier frequency to each animal. An approach to this problem is to use several beacons on the same frequency, but use different pulse rates for each beacon. One then has to develop a system to unscramble the overlayed pulses so that individual animals can be identified. Cupal, 1977, developed a novel system for doing this by using an RCA COSMAC 1801 microprocessor. The system program is developed around time differences between received pulses to separate and identify individual beacons. Although a certain amount of hardware is required, the success of the system is in the software, which is described in the reference cited.

11.7. ANTENNA CONSIDERATIONS

The single most difficult problem in the design of a biotelemetry system is the transmitting antenna. Receiving antennas are normally some conventional design, cut for the proper wavelength, which has both suitable gain (directivity) and polarization for the telemetry system in use. General design considerations can be found in references such as Jordan and Balmain, 1968. The transmitting antenna, however, presents some very special problems, one of which is the layered medium problem. If we consider a biotelemeter for transmitting information from inside of a free-swimming fish, there are two interfaces that affect transmission: 1. The fish–water interface, and 2. The water–air interface.

Let us now examine the water–air interface problem. This has been discussed by Lindsay *et al.*, 1977, and Weeks *et al.*, 1977. The maximum physical size for a biotelemetry transmitting antenna is limited by the physical size of the subject animal. Antenna size also depends upon the dielectric constant of the medium through which radio propagation occurs. The dielectric constant of water is approximately 80; that of air is one. From antenna theory, this means that the size of an antenna in water is a factor of the square root of 80 (or one-ninth) smaller than its size in air.

The radio frequency at which the system operates best depends upon the electrical conductivity of the water. Alpine lake water derived from snow melt ordinarily has low conductivity. Lowland lake and river water generally contains large quantities of dissolved salts and has high conductivity. In addition to dielectric constant, radio propagation depends upon the electrical conductivity of the propagation medium. Generally speaking, the higher the conductivity, the lower the frequency (or the longer the wavelength) must be for effective radio transmission. Since antenna size is directly proportional to wavelength, there are several tradeoffs that must be made between animal size, antenna size, and telemeter operating frequency. It may not be possible to transmit effectively from aquatic subjects swimming in highly conductive water. Salt water produces extreme problems and frequently ultrasonic systems are employed rather than radio techniques. Ultrasound propagates as longitudinal waves and is more suitable for propagation of energy through high-loss media, while radio waves are transverse. Both the transmitting and receiving antennas are immersed in ultrasonic systems.

To gather some idea of dimensions, let us look at the case of radio transmission from a freshwater game fish. A reasonable size telemeter package has a major dimension of 10 cm. Normally, half-wavelength antennas are used. A half wavelength of 10 cm in water corresponds to 90 cm in air, with an associated frequency around 166 MHz. Radio waves at this frequency will penetrate through several meters of low conductivity water. As conductivity increases, penetration distance decreases. For good conductors such as salt water, penetration depth δ (meters) is given approximately by the relation:

$$\delta \sim \sqrt{\frac{2}{\omega\mu\sigma}}$$

where ω is the radian frequency of propagation, μ is the 4×10^{-7} H/m, and σ is the conductivity in Siemens.

When the conductivity of the medium is very low,

$$\delta \sim \frac{2}{\sigma}\sqrt{\frac{\epsilon}{\mu}}$$

For the general case (neither approximation valid), that is, where $\sigma \sim \omega\epsilon$

$$\delta = \left[\omega\sqrt{\mu\epsilon/2}\sqrt{\sqrt{1 + (\sigma/\omega\epsilon)^2} - 1}\right]^{-1}$$

Thus not only the electrical properties of the medium (σ, ϵ), but also the frequency of the radio waves ($\omega = 2\pi f$) affect the depth of penetration.

Another consideration is antenna input impedance, since this affects the design of the transmitter output stage in the telemetry unit. A half-wave dipole antenna in air has an input impedance of approximately 72 Ω. In water, this is reduced by the $\sqrt{80}$ to approximately 8 Ω. Thus although antenna size is reduced in water, which is advantageous, antenna input impedance is also reduced, which is not advantageous because of loading affects upon the transmitter. The actual problem is matching the impedance of the transmitter output circuit to the low input impedance of the antenna. The matching networks required for this add electrical losses to the system and reduce efficiency. Matching can certainly be obtained, but at the expense of high efficiency operation.

There are several parameters that must be considered when transmitting radio waves through the water–air interface, and we will now examine these one at a time. We first define the plane of incidence as that plane in which both the ray path and the normal vector to the water–air interface are contained. We will assume linear polarization of the antennas, as opposed to circular or elliptical polarization. Normally, the antenna on a fish would be aligned along the principal axis of the fish, and hence either parallel or perpendicular to the plane of incidence, which produces either horizontal or vertical polarization of the antenna. Energy transmitted from the fish to a receiving antenna above the water–air interface depends upon the vector components of the electric field, which are both parallel and vertical to the plane of incidence. Figure 11.16 illustrates radio-wave ray paths for this situation and the corresponding polarizations of the electric field in air. In water, the angle of incidence θ_1 is generally from 0 to about 6.4°. For this range of the angle of incidence, the angle of the transmitted wave in air, θ_2, ranges from 0 to 90°. If the angle of incidence increases beyond 6.4°, the energy is reflected back into the water by the interface.

The amount of power transmitted through the interface into the air is a function of polarization and the angle of incidence. Figure 11.17 represents this situation. Complete power transmission is unity in the diagram. Since an aquatic animal may be randomly oriented, the signal at the receiving antenna may contain both horizontally and vertically polarized components. To accommodate this condition, the receiving antenna should be circularly polarized. If tracking is to be done by aircraft, the antenna should be mounted beneath the craft.

Analysis for the animal–water interface is very complex, since TEM waves are not involved, but rather the antenna near field pattern

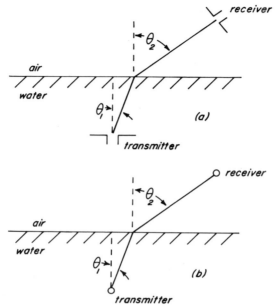

FIG. 11.16   Ray paths for radio wave propagation across water–air interface (after Lindsay *et al.*, 1977; Weeks *et al.*, 1977). (a) Ray path for polarization parallel to the plane of incidence, $0 \leqslant |\theta_1| \leqslant 6.4°$; (b) Ray path for polarization perpendicular to the plane of incidence, $0 \leqslant |\theta_1| \leqslant 6.4°$.

analogous to the Fresnel zone in optics. In addition, there are varying electrical properties because of the layered nature of animal tissue (see Schwan, 1957, for a discussion of layered tissue). Generally one approaches this problem experimentally rather than analytically. One approach to this problem is the use of double telemeters as described in the next section.

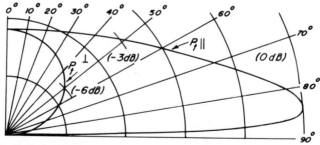

FIG. 11.17   Relative power transmitted to air region vs angle of elevation for perpendicular and parallel polarization (after Lindsay *et al.*, 1977; Weeks *et al.*, 1977).

11.8.   TWO-STAGE TELEMETRY SYSTEMS

One drawback with implanted telemetry transmitters is the finite lifetime of the batteries. When transmission occurs from inside of an animal and the unit must have a range of more than a few meters, the current drain on the batteries will be large because of the interface and media effects discussed in section 11.7. Replacement of the telemeter package requires capturing the animal and surgery. Telemeter package size limitations preclude using very high ampere–hour capacity batteries.

An approach to this problem uses two telemeters. A very low power unit is implanted in the animal. It generally uses a flat loop or coil antenna placed subcutaneously. A similar surface-attached antenna receives the signal, by inductive coupling, transmitted through the skin. This signal is then retransmitted by a high power telemeter externally attached to the animal. This unit may be contained in some sort of harness or neck collar. The main signal may be simply a rebroadcast of the low-level signal using an amplifier and transmitting antenna, or the low-level signal may be frequency multiplied and then retransmitted. This method has several advantages. The implanted unit has a long lifetime because of low power output and reduced battery drain. The external unit can be replaced relatively easily when the batteries begin to run down.

In some respects, the radio-frequency and inductive cardiac pacemakers (Glenn, 1964) are the inverse of this method. In this case, a passive receiver is implanted within the body and energy is transmitted from an external power unit to a subcutaneous receiving coil (antenna).

11.9.   MODULATION TECHNIQUES

In terms of requirements for both the transmitter and receiver associated with a telemetry system, it is generally simpler to use amplitude modulation (AM) techniques. When transmission is over relatively short distances, interference, noise, and reduction in signal-to-noise ratio (S/N) are not usually significant. Less channel bandwidth is required in AM over other techniques and this is an advantage.

When multiple channels of information are to be transmitted over a single-carrier frequency, as depicted diagrammatically in Fig. 11.18, then frequency modulation (FM) transmission is required. Modulation and demodulation is more difficult and channel bandwidth is increased.

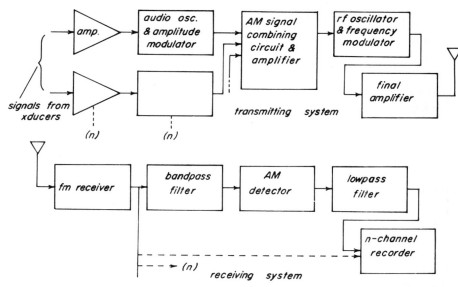

FIG. 11.18   FM system for multiple signal (multiplex) transmission.

FM systems are frequently specified because of the better signal fidelity (frequency response) which can be obtained. This is true when one is dealing with wide spectrum signals such as are found in audio systems. We must recognize, however, that most biological signals are relatively low frequency and narrow in spectrum. Normally the ECG, for example, is not considered to have significant spectral components greater than 40 Hz. When these factors are taken into consideration, with careful design, AM systems can be made to perform as well as FM systems if only one channel of information is to be transmitted.

Pulse modulation finds application in beacons and certain other systems, as has been indicated in previous sections. The Appendix (below) illustrates the spectral characteristics of simple AM and FM systems.

Appendix

Designate all modulating waveforms by $f_m(t)$ and all carrier waveforms by $f_c(t)$. It is assumed that the modulating signal varies very slowly relative to the carrier frequency, otherwise we cannot describe an envelope function. Thus $\omega_m \ll \omega_c$, where ω_m and ω_c are, respectively, the instantaneous radian frequencies associated with the modulation and the carrier.

Let $f_c(t) = A \cos \omega_c t$

Amplitude Modulation

An AM waveform is described by

$$f_{AM}(t) = [1 + mf_m(t)]f_c(t)$$
$$= A[1 + mf_m(t)] \cos \omega_c t$$

If the modulating signal is a pure tone, then

$$f_m(t) = \cos \omega_m t \text{ (normalizing to unit amplitude)}$$

and

$$f_{AM}(t) = A[1 + m \cos \omega_m t] \cos \omega_c t$$
$$= A \cos \omega_c t + (A/2)m \cos (\omega_c + \omega_m)t + (A/2)m \cos (\omega_c - \omega_m)t$$

This last expression shows that the AM signal consists of a carrier at frequency ω_c and two sidebands at frequency $(\omega_c + \omega_m)$ and $(\omega_c - \omega_m)$. The modulation index m varies between 0 and 1. If $m = 0$, there is no modulation. If $m = 1$, 100% modulation exists. Normally m is set to be close to one.

$$f_{AM}(t) = A \cos \omega_c t \qquad m = 0, \text{ no modulation}$$
$$f_{AM}(t) = A[1 + \cos \omega_m t] \cos \omega_c t \qquad m = 1, 100\% \text{ modulation}$$

Frequency Modulation

An FM waveform is described by

$$f_{FM}(t) = A \cos (\omega_c t + \beta \sin \omega_m t)$$

in which the modulation index β is defined by

$$\beta = \Delta\omega/\omega_m = \Delta f/f_m$$

β is the maximum phase shift of the carrier and is defined as the ratio of the deviation $\Delta\omega$, or Δf, of the carrier to the modulation frequency ω_m, or f_m. Generally β is quite small. The FM waveform has a complex spectral expansion and is given by

$$f_{FM}(t) = J_0(\beta) \cos \omega_c t + \sum_{n=1}^{\infty} \{(-1)^n J_n(\beta)[\cos (\omega_c - n\omega_m)t + (-1)^n \cos (\omega_c + n\omega_m)t]\}$$

where $J_n(\beta)$, $(n = 0, 1, 2 \ldots)$ is the nth order Bessel function of the first kind.

11.10. SOME RECEIVER CONSIDERATIONS

The treatment of receiving antennas, by choice, has been rather cursory. As previously noted, they must be cut to the correct wavelength, have appropriate gain, and be correctly polarized to maximize the signal received. The input stage of most modern receivers suitable to telemetry applications uses an RCA dual-input-gate MOSFET. This produces a very low noise receiver front end. With such receivers and some signal modulation, such as the signal from a beacon, 0.01–0.03 μV at the receiving antenna terminals is sufficient for field tracking.

It must be remembered that the human ear is one of the best detectors, filters, and discriminators available. It is possible to separate signals by ear that are nearly impossible to separate by normal electronic techniques. This is responsible for the 0.01–0.03 μV signal level. When carrier signals are modulated with one or more analog channels, larger input signals are generally required so that the demodulated channels have adequate S/N ratios.

11.11.   SUMMARY

The intent of this chapter has been to present a wide-ranging review of techniques used in biotelemetry systems with some discussion of state of the art methods. In a single chapter of a general bioinstrumentation text, it is not possible to discuss all of the design considerations for transmitters and receivers. Major points have been treated and the interested reader is directed to the literature citations at the end of this chapter.

11.12.   REFERENCES

Caceres, C. A., ed., 1968, *Biomedical Telemetry*, Academic Press, New York.
Cupal, J. J., 1977, "A Transmitter Identifier for Use with Wildlife Biotelemetry," *Biomed. Sci. Instr.* **13**, 13–17.
Ferris, C. D., and S. D. Hakes, 1975, "A Dry-Electrode Electrocardiograph System," *Biomed. Sci. Instr.* **11**, 15–19.
Glenn, W. W. L., 1964, "Cardiac Pacemakers," *Ann. N.Y. Acad. Sci.* **111**, 813–1122.
Jordan, E. C., and K. G. Balmain, 1968, *Electromagnetic Waves and Radiating Systems*, Prentice-Hall, Englewood Cliffs, N.J.
Lindsay, J. E., F. M. Long, and R. W. Weeks, 1977, "A Practical Look at Antenna and Propagation Requirements in Biotelemetry Systems for Freshwater Fish, *Proc. IEEE Ant. Prop. Soc. Int'l. Symp.*, 132–135.
Lonsdale, E. M., 1969, "Radiotelemetry as a Tool in the Study of Freshwater Fish," *Conf. Rec. Sixth Ann. Rocky Mtn. Bioengrg. Symp.*, Laramie, Wyo., pp. 40–45.
McAleenan, R. N., R. W. Weeks, and F. M. Long, 1976, Computer Aided Biotelemetry System Applied to Free Swimming Fish, *Biomed. Sci. Instr.* **12**, 29–32.

MacKay, R. S., 1959, "Radio Telemetering from within the Human Body," *Trans. IRE Med. Electron.*, ME-6(2), 100–105.

MacKay, R. S., 1960, "Endoradiosondes: Further Notes," *Trans. IRE Med. Electron.*, ME-7(2), 67–73.

MacKay, R. S., 1968, *Bio-medical Telemetry: Sensing and Transmitting Biological Information from Animals and Man*, Wiley, New York.

Schwan, H. P., 1957, in *Advances in Biological and Medical Physics V*, J. H. Lawrence and C. A. Tobias, eds., Academic Press, New York.

Slagle, A. K., 1965, "Designing Systems for the Field," *Bioscience*, 15(2), 109–112.

Weeks, R. W., F. M. Long, and J. J. Cupal, 1976," Radio-recoverable Tranquilizing Dart," *Biomed. Sci. Instrum.*, 12, 33–36.

Weeks, R. W., F. M. Long, J. E. Lindsay, R. Baily, D. Patula, and M. Green, 1977, Fish Tracking from the Air, *Proc. 1st Int'l. Conf. on Wildlife Telemetry*, 63–69.

NASA *Tech. Briefs*, 1975–1977.

Section 6

PRACTICAL MATTERS

12

Some Practical Matters

There are numerous practical matters associated with the design, purchase, and use of instrumentation systems that need to be considered. These are usually learned "on the job" rather than in a formal text presentation, and some of the more important items are discussed in this final chapter.

12.1. GROUNDING

There are several reasons for taking pains to install proper grounding in instrumentation systems. A major one with legal implications is user/patient safety. Beyond that, improper grounding, as well as producing electrical hazards, can result in excessive noise and 60 Hz pickup in high-impedance low-signal-level portions of a system, with concomitant degradation of S/N and erratic system performance.

12.1.1. Safety Aspects

The NFPA publishes the National Electrical Code which delineates the basic requirements for electrical service in residential, industrial, health-care facilities, and other installations. Instruments, although not covered *per se* in the Code, should be designed to be compatible and comply with the regulations and requirements cited. A basic requirement is that a 3-wire power cord should be used in instrument design. Most instruments operate on 60 Hz at a nominal 120 V_{rms} AC. X-Ray units and some other instruments may operate on 220 (208) or 440 V AC, but these are in the minority. The basic requirement is that electrical shock hazard be eliminated.

In the 3-wire power cord configuration, the ground wire (circular prong on plug) is connected to the metal cover of the instrument or chassis frame. If a fault occurs such that the "hot" wire in the 120 V supply contacts the cover or chassis of the instrument (such as by a partial break in the cable insulation), then a short circuit occurs since the case is grounded. This will cause a fuse to blow or a circuit breaker to interrupt in the power distribution circuit, but the instrument user is protected from electrical shock. This, of course, assumes that the power distribution system into which the instrument is connected has been correctly designed and installed. When a system is properly installed the receptacle should exhibit the voltages indicated in Fig. 12.1. In the same figure, the design of a connection fault tester is shown.

Table 12.1 delineates electric shock levels and the associated physiological manifestations.

Electric shock can be divided into two categories: microshock and macroshock. Microshock describes the internal shock that can occur during certain diagnostic (cardiac catheterization) and surgical procedures. It generally results from an improperly grounded transducer

Table 12.1

Physiological Responses to Various Levels of Electric Shock

Current level	Effect
	Microshock
10–20 μA	Ventricular fibrillation
	Macroshock
1 mA	Sensation
5 mA	Upper limit to harmless level
10–20 mA	"Let go" threshold; flexor muscles stronger than extensor muscles
30–40 mA	Tetany; sustained muscle contraction and cramping
50–70 mA	Pain, exhaustion, fainting, irreversible nerve damage
100 mA	Fibrillation and death, if current passes through body trunk
> 100 mA	Fibrillation, amnesia, burns, severe electrolysis at contact sites
> 5 A	Little likelihood for survival

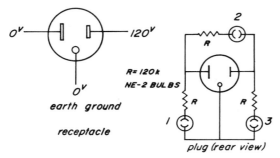

FIG. 12.1   Receptacle and tester for 120 V AC system. Bulbs 2 and 3 should light if system is correctly wired.

probe being placed into a patient while the patient is externally grounded through an electrocardiograph recorder, or perhaps by the patient touching the metal frame of the examination table upon which he or she is lying. In situations of this nature, there are direct electrical paths to the heart and very small currents can produce ventricular fibrillation.

Macroshock refers to simultaneous body surface contact with a "hot" wire and a grounded object. Higher currents are involved since direct electrical paths to the heart are less evident. The susceptibility of the ventricles to fibrillation depends upon both the frequency and the energy content of the excitation signal. By an unfortunate turn of events, 60 Hz is about the most lethal from a frequency point of view. The electrical impedance of the body to 60 Hz is low.

To protect patients who are connected to diagnostic or therapeutic equipment from shock, isolation units have been developed which, when operating correctly, isolate the patient from the equipment and the power system. Normal circuit breakers and fuses operate at relatively high currents and would not interrupt a circuit for a microshock fault current in the 10–20 μA range.

12.1.2.   Patient Isolators

There are several methods by which a patient can be isolated from a monitor. The simplest is a current limiter in series with each active (ungrounded) patient lead. Figure 12.2 illustrates one basic technique and the volt–ampere characteristic of the diode pair used in this system. R_m and R_p are series current-limiting resistors. If R_p is omitted, faults originating from the patient side are unprotected; if R_m is omitted, no protection occurs for faults originating on the monitor side. In operation, if a large voltage (regardless of polarity) appears between point A

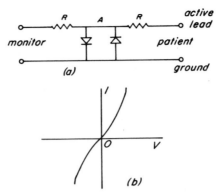

FIG. 12.2   Wiring diagram (a) and volt–ampere characteristics (b) for patient-monitor lead fault protector.

and ground, one of the diodes conducts and limits this voltage to the forward diode voltage (~ 0.7 V for silicon diodes). The series resistors R_m and R_p limit current and, in theory, prevent diode burnout. The bilateral protection system shown in the figure protects the patient from monitor faults, and it also protects the monitor input from high voltages such as occur in the use of defibrillators applied to the patient. For use with a conventional four-wire ECG lead system, the connections are as shown in Fig. 12.3. One commercially available unit (Ohmic Instruments Co.) uses 1N4001 diodes and 20 kΩ, 2 W film resistors for R_m; R_p is not used. When balanced lead systems are used, then the circuit diagrammed in Fig. 12.4 applies.

Series limiting devices are fine as long as a component does not burn out. With *external connections* to the patient and the fault voltage limited to 0.7 V, the maximum fault current delivered to the patient would be 140 μA assuming 5 kΩ impedance between patient electrodes and $R_p = 0$. For $R_p = 20$ kΩ, this current would drop to 28 μA for

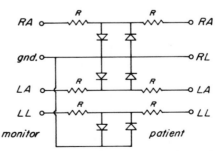

FIG. 12.3   Wiring of patient isolation system for four-wire ECG system.

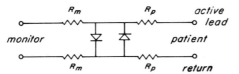

FIG. 12.4   Symmetric isolator for two-active-lead system.

the circuit of Fig. 12.2. Frequently low-current fuses are inserted in series with the resistors for added protection should a long-term fault occur. One limitation in the schemes presented in Fig. 12.2 to Fig. 12.4 is a reduction in desired signal level because of the series resistors.

Loss of desired signal may be reduced by using field-effect diodes in the current limiting circuit, as shown in Fig. 12.5. Two of these devices connected in opposition provide a linear voltage–current characteristic before current limiting. Thus less signal loss occurs since R_m and R_p have lower values than in the previous circuits. The diodes R_m and R_p must be selected so that sufficient current limiting occurs to protect the patient. Let us consider a sample calculation where we require monitor protection during 700 V defibrillator pulses. Assume 5 s pulse intervals and pulse durations of 0.1 s (these values are characteristic of some AC defibrillators). For convenience, we choose $R_m = R_p$.

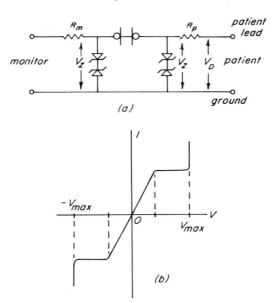

FIG. 12.5   Zener diode–field effect diode isolator system (a) and its volt-ampere characteristics (b).

We select zener diodes for 2 V regulation in compliance with specifications for the monitor. If we assume zeners with a power rating of 250 mW, the maximum zener current is 125 mA and the normal forward diode (surge) current is

$$I = 0.25/0.7 = 357 \text{ mA}$$

assuming 0.7 V forward drop across the diode. If we neglect the 0.7 V drop across the forward-biased diode

$$R_m = R_p \sim \frac{V_D - V_z}{I_z} = \frac{700 - 2}{0.125} = 5584 \ \Omega$$

$$= 5600 \ \Omega \text{ (standard value)}$$

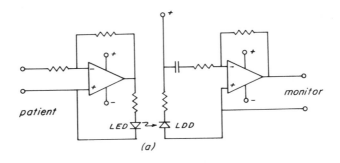

(a)

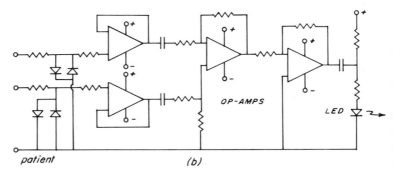

(b)

FIG. 12.6 (a) Simple optical isolator for patient lead system. (b) More complex optical isolator featuring series resistance and shunt-diode input protection. The LED is forward biased to maintain linear operation with applied signal. The input follower op-amps should be high-input-impedance FET types. This isolator will serve for an ECG channel with two active leads and a patient ground (reference) lead. A somewhat more sophisticated system has been proposed by Holsinger and Kempner, 1971. Holmer, 1974, has suggested an acoustic isolator.

Power dissipated in R_m and R_p:

$$P \sim \frac{\tau(V_D - V_z)^2}{5600} = \frac{(698)^2(0.1/5)}{5600} = 1.74 \text{ W}$$

where τ = pulse duration/repetition rate. The surge current in the zeners is

$$I_s \sim \frac{V_D - V_z}{5600} = \frac{700 - 2}{5600} = 124.6 \text{ mA}$$

The field effect diodes typically have internal impedances of the order of 1000 Ω. Constant current rating is chosen not to exceed the safe limit for a human subject. The total series resistance in the circuit is $2 \times 5600 + 1000 = 12,200$ Ω, which allows an ECG signal current (for a 1.5 mV peak R-wave) of 0.12 μA $= 120$ nA.

Although the sorts of devices discussed above do protect the patient (or monitor), they do not isolate the patient completely from the power source. Several alternative techniques have been suggested to ameliorate this difficulty. One is to use transformer coupling between patient and monitor. Generally transformers are equivalent to relatively narrow-band band-pass filters. Their insertion into patient leads would introduce excessive signal distortion. Another isolation technique is telemetry, but this is expensive and subject to noise and other problems, as discussed in the last chapter. A current method is the use of optical isolators. These provide complete isolation at reasonable cost. A typical system is shown in Fig. 12.6.

12.1.3.   Ground Monitors

Two types of ground monitoring systems are in common use. One type is used to check the continuity of the ground system, and the other type to check for faults. *The reader should carefully review the Legal Note at the end of this section.* A passive four-wire continuity checking system is shown in Fig. 12.7. A redundant ground wire is used and the inputs to the operational amplifier are tied to the two ground wires. With proper ground continuity, the two inputs to the amplifier have the same potential. With a break in the grounding system, the input potentials are unbalanced and the output voltage of operational amplifier increases above 5 mV, which in turn triggers the alarm.

The active system shown in Fig. 12.8 has also been proposed for monitoring ground continuity. A high-frequency low-voltage current

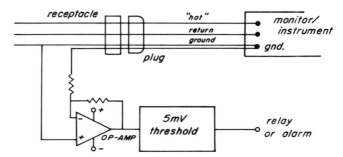

FIG. 12.7   Four-wire passive system for monitoring ground wire continuity.

signal is injected in series with one of the redundant ground wires. The current in the other wire should be the same so that there is no potential difference across the input terminals of the operational amplifier with respect to the injected signal. Loss of ground continuity produces an imbalance in the potentials applied to the op-amp, which drives the output high to trigger the threshold circuit and alarm. This system has several drawbacks. One is that a small current must be injected in the normally zero-current true ground. This current may be detected as a fault current by other equipment. Another problem relates to the use of a high frequency signal (> 60 Hz) which may produce interference in patient signals.

A number of schemes have been proposed for detecting ground faults, as opposed to loss of ground continuity. A fault is a current in the ground system that results from a short circuit or partial short circuit to ground from one of the "hot" wires in the power distribution system. Figure 12.9 depicts one simple system for detecting ground faults when the local power-distribution system is isolated by an isolation transformer from the main power bus. Use of isolation transformers is generally recommended in critical care areas and for sensitive

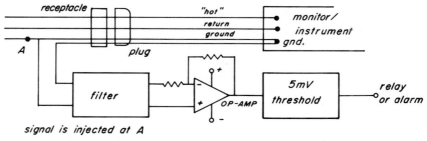

FIG. 12.8   Four-wire active system for monitoring ground wire continuity.

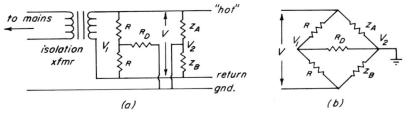

FIG. 12.9   Wheatstone bridge fault detection system (a) and equivalent Wheatstone bridge (b).

instruments. The circuit is a Wheatstone bridge. Z_A and Z_B represent fault impedances, and the resistors R are always connected. In correct operation of the power system, $V_1 = V_2 = 0$, and $Z_A = Z_B \rightarrow \infty$. R_D is the internal resistance of the fault detector, which may be either a current-sensing or voltage-sensing device. If a fault occurs, then either Z_A or $Z_B \neq 0$ and $V_1 \neq 0$, which establishes a current in R_D. If this current is interpreted as a voltage drop across a true resistor R_D, then a suitable detector is the system shown in Fig. 12.10. The output from the op-amp can be used to actuate an alarm or to control a relay, which in turn interrupts the power source. The detection system has one drawback and that is a "blind spot" when $Z_A = Z_B$. In this case of a balanced fault, the bridge remains balanced ($V_1 = V_2$) and the fault is not detected.

In order to achieve initial bridge balance in this system, it may be necessary to shunt the two resistors (R) by small capacitors. It must be remembered that in an AC system, bridge balance is defined by:

$$V_1 \underline{/\theta_1} = V_2 \underline{/\theta_2}$$

or

$$V_1 = V_2 \quad \text{and} \quad \theta_1 = \theta_2$$

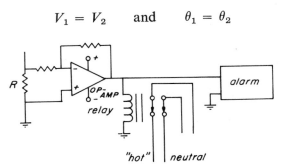

FIG. 12.10   Fault current monitor. See legal note in text regarding insertion of a resistance in a ground lead.

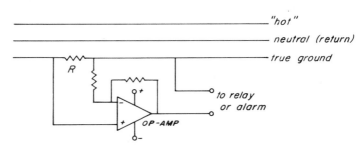

FIG. 12.11   Fault current monitor using resistor inserted in true ground wire (not now a legally approved method—see text and legal note).

Thus the detector will respond to both resistive faults (full or partial short circuits) and high capacitive leakage faults ($Z = 1/j\omega C$).

For a power system that is not isolated through a transformer, a fault detector can be instrumented as shown in Fig. 12.11. The design has one serious drawback in that a small resistor must be inserted in series with the normally zero-resistance true ground wire. A fault to ground will establish a current in and concomitant voltage drop across R. This voltage is sensed by the op-amp, whose output is used to control a relay or alarm as in the system of Fig. 12.11. This circuit, using a *practical* op-amp, will also act as a ground continuity detector for an open ground to the left of R as the diagram is drawn. If an open circuit occurs, then the input to the op-amp floats. The input is referenced to power line ground, which is the right side of the diagram as drawn. The system will *not* operate as a ground continuity detector for open circuits to the right of R. In the case of *ideal* op-amps, the system cannot be used as a continuity detector because the two input terminals would float at the same potential, and there would be no potential difference to be amplified for a break in continuity to the left of R.

Legal Note

This circuit has been included as a historical note. At the time of writing there is legislation pending concerning the design of fault detectors. *The system shown in Fig. 12.11 is not legal since insertion of R into the ground line could result in raising the potential of the "true" ground to dangerous levels. One involved in the design and use of fault detectors should consult with appropriate authorities concerning design and application legality and limitation.*

12.1.4.   Ground Potential Differences

Because of the nonzero electrical resistance of copper or aluminum conductors, a ground is not always ground. Multistoried buildings

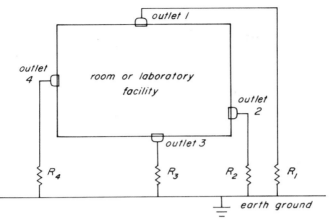

FIG. 12.12 Demonstration of unequal ground wire resistances because of differences in the physical lengths of the grounding wires. This leads to differences in ground potentials and to ground loops. Note: $R_1 \neq R_2 \neq R_3 \neq R_4$.

are usually wired with vertical runs of wire. Thus in any given room, each electrical outlet may be on a different circuit, as indicated in Fig. 12.12. This means, in general, that the lengths of the grounding wires to the point where they are tied into the distribution system differ. If fault currents are present in these ground wires (third wires), then because of the different lengths and different resistances, there are potential differences between the "grounds" in the same room. These potential differences can be sufficient to produce microshock hazards. To prevent this condition in critical care areas, a common ground bus should be installed and all grounds tied to it; this is a heavy copper bar. This varying ground problem can also produce ground loops, as described below.

12.1.5. Ground Loops

Ground loops develop when ground return wires are connected to different points in the grounding system. Because of varying potential drops, as noted above, with varying lengths of the grounding conductors, different points of a circuit may have different ground potentials. This situation also occurs when different instruments are interconnected in a measurement system. Here the problem is aggravated by the third-wire common ground in the power cord. In most instruments, signal ground is also instrument case ground; but owing to safety requirements, the case ground is connected to the third wire of the power cord. Because of different values of resistance in the ground wires, numerous potential differences can develop when instruments are plugged into different outlets, as was demonstrated in Fig. 12.12.

Ground loops are not so much a safety problem (because the potential differences are usually very small), but rather a signal interference problem. Since 60 Hz potentials are involved, the ground difference potentials can be additive to desired low-level signals in the instrumentation system. This results in 60 Hz or "hum" modulation of signals and degradation of the S/N ratio. The solution to this problem is to connect the power system ground at a common point. Use of a multi-outlet receptacle box (provided it has sufficient current capacity) will usually solve the problem. In some cases, it may be necessary to disconnect the third-wire grounds, although this practice can produce a safety hazard, or to isolate signal ground from power system ground. Within a given instrument, grounds should be returned to a common point or grounding bus to avoid loops.

To avoid external ground loops, redundant ground paths or connections should be eliminated. System or instrument grounds should be made at a common point. One tends to forget that ground wires do have resistance across which potential drops can occur when fault or leakage currents exist. Because of capacitive effects, there is always some leakage current in a 60 Hz system.

12.2. ELECTRICAL SHIELDING

High impedance circuits act as antennas and electrical energy may couple into them inductively or capacitively. Undesirable signals include 60 Hz pickup, switching and commutator noise from electric motors and machinery, radio frequency interference from fluorescent lighting fixtures, diathermy apparatus, and arc-welding equipment. With the wide use of semiconductor and integrated circuits, which generally operate at low impedance levels, pickup of spurious signals is much less a problem than it was with high-impedance vacuum-tube circuits. In instrumentation systems, however, there are some transducers that characteristically operate at high impedance levels. These include microelectrodes, which operate into electrometer-type amplifiers and reverse-biased photodetectors, which utilize high-value load resistances. In both of these cases, the signals involved are very low level. Generally some sort of electromagnetic shielding is required when the combination of low signal levels and high impedances occurs.

Shielding generally involves placing the circuit in a copper (best) or aluminum enclosure that is properly grounded at one point to the instrumentation system ground. The system as a whole should be grounded at one point to a *reliable* earth ground. Generally speaking, water or steam pipes are not reliable earth grounds because electrically insulating flanges and gaskets may be used in the plumbing system to reduce electrolytic effects. To provide a satisfactory ground, a solid copper rod (at least 3/4″ in diameter) should be driven 6–8 ft into the earth, and the grounding system wire (at least #10) connected to it. The soil around the rod should be kept moistened with saline solution. Within the shield, all ground connections should be made at a single point. The shield should enclose entirely the sensitive circuit with any connections to other circuits brought through small holes in the shield.

In some cases, it is necessary to use shielded connecting cables. If the system has a continuous common ground, *the shields should be grounded at one cable end only when low-level signals are being processed.* The reason for this is to avoid ground loops through the cable shield. In such applications, the cable should be a shield and not a return wire. To reduce pickup, the signal wire and its return can be twisted together to form a twisted pair. The pair in turn can be enclosed within a flexible braided shielding sheath. Such cables are commercially available.

In sensitive circuits, component and grounding leads should be kept as short in length as possible. This is usually easily accomplished with printed circuit board layouts. High-signal-level and power supply leads should not be laid out next to low-signal-level conductors. Where it is necessary to have high and low level circuits in close proximity, they should be arranged at right angles to one another.

Shields should be constructed of solid sheet metal, but in some applications (such as microelectrode work) it may be necessary to see the circuit. In these cases, it is possible to use copper screening that is soldered completely along any joining seams.

12.3.   ELECTROSURGERY APPARATUS (ESA)

Not all electrical hazards originate from 60 Hz equipment. Electrosurgery apparatus may be the source of severe burns in both surgical patients and the physician operator. These devices utilize high-frequency electric currents to cut, coagulate, and desiccate tissues. They normally operate in the range from 0.3 to 5.0 MHz, which minimizes muscle stimulation (electric shock), but yet provides the desired surgical effects.

The basic system consists of a high-power high-radio-frequency generator, the cutting electrode, and a dispersive electrode that is patient ground. The dispersive electrode provides the rf signal return path connection to the generator. Power output may be as much as 500 W although 50–100 W is normal for open air cutting. Higher powers may be required when tissues are irrigated with fluids. A constant-amplitude sinusoidal waveform is generally used for cutting, but pulse-modulated sinusoids are used for cautery. Combination waveforms are also used in some applications.

Figure 12.13(a) illustrates the basic system. Neither the rf generator

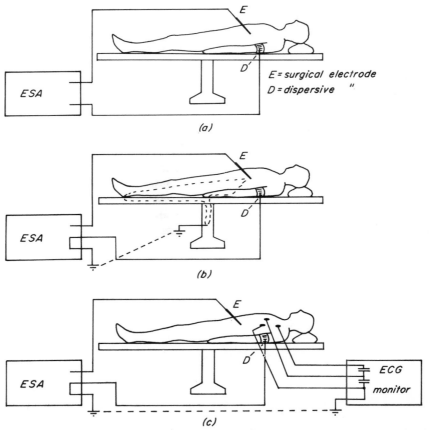

FIG. 12.13 Burn hazards associated with electrosurgical units. (a) Normal system operation; (b) Burn hazard resulting from common ground between ESA and operating table; (c) Burn hazard at monitoring electrodes resulting from common ground between ESA and ECG and stray capacitance.

(ESA) nor the operating table is grounded. The dispersive electrode may be a buttocks plate, or a flexible plate wrapped around the thigh or upper arm (as shown in the figure). It is typically a flexible metal-foil plate, on the order of 100 cm² or more in area. Electrical continuity between the plate and the patient is maintained by a conducting gel. There are several proprietary formulae for gels. In normal operation, the rf current passes from the high side of the generator through the connecting cable to the surgical electrode, though the patient to the dispersive electrode, and back to the low side of the generator via the connecting cable from the dispersive electrode. The surgical electrode has minimal area, so that the tissue damage is confined to a very small region surrounding this electrode. In effect, the electrode produces a radio-frequency burn. Because of its large area, and consequently low rf current density, the dispersive electrode prevents burning at the patient ground site, provided that electrical contact is maintained over the entire electrode surface. Poor contact between the patient and the dispersive electrode can result in rf "hot spots" and subsequent burning of the patient.

Because of the high frequency of electrosurgical currents, stray capacitance currents are important in addition to ohmic currents. The magnitude of capacitive impedance (capacitive reactance) is given by $1/\omega C = 1/(2\pi fC)$. Thus the higher the frequency, the lower the impedance of a given capacitor. Figure 12.13(b) illustrates a burn hazard situation. In this case, both the operating table and the electrosurgery unit apparatus (ESA) are grounded. The patient is touching the grounded table at the left heel and left hand. Body tissues have both resistive and capacitive properties. As shown by the dotted lines in the figure, capacitive coupling exists to the heel and hand contact sites, resulting in burns to the patient at these sites.

Figure 12.13(c) illustrates a hazard associated with the use of an ECG monitor in conjunction with an ESA. Capacitance is again the problem. Either capacitance between wires in the patient connecting cable, or stray capacitance within the monitor (or both) may be factors. The rf energy from the surgical electrode couples to the voltage sensing electrodes and is returned to ground through the stray capacitance. This produces burns at the monitoring electrode sites. The smaller the monitoring electrodes, the more severe the burns. This situation may be prevented by inserting rf choke coils in series with the monitoring electrodes. The rf chokes will pass DC and low frequency currents, but not the high frequency currents associated with the ESA. Monitoring electrode cables with built-in rf chokes are now commercially available (NDM Corp., Dayton, Ohio).

Other burn hazards from electrosurgical units also occur; the principal ones have been described here. As a general rule, the ESA, the operating table, and any monitoring equipment should be isolated from one another. One should also refer to the National Fire Protection Association publication: *High Frequency Electrical Equipment in Hospitals—1970*, NFPA 76 CM, 1970.

12.4.   EQUIPMENT SELECTION

Many instrumentation systems are not designed "from scratch," but are assembled from available manufactured items, such as transducers, a wide variety of amplifiers, tape transports, chart recorders, and oscilloscopes. In selecting ready-made components to incorporate into a system, one must weigh many factors. Some of the items to consider are:

Cost	Impedance levels
Reliability	Signal sensitivity
Compatibility with other components	Ease of use
	Mechanical design and ruggedness
Maintenance requirements and ease in servicing	Frequency response
	Distortion characteristics
Ease and stability of calibration	Portability
Accuracy (absolute value)	Versatility (is the unit special-
Precision (repeatability of a given measurement)	purpose or can it be modified to perform other functions?)
Power system or bias requirements	Apparent nonlinearities
Susceptibility to spurious signals	Risetimes and time constants
	Common-mode signal rejection

In terms of specifications, two very similar appearing instruments may have substantially different costs. It is then up to the purchaser to determine why. Price and quality are not always linearly related. Frequently, price differences relate to rather subtle differences in design features, which may be nothing more than pushbutton switches versus rotary switches. One can wind up paying a high cost for slight convenience differences. Frequently, physical size of the device will affect cost. Miniaturized equipment is often more expensive. Other factors are output options, such as a digital output available as a plug connection in addition to the normal analog display associated with the device.

Frequency response, distortion figures, and risetime features often produce substantial cost differences in similar pieces of equipment. The instrument user may need to compare very carefully application requirements versus the specifications of available equipment. One should look for versatility, but on the other hand, not purchase more than is really necessary to do the job, unless one enjoys the enviable position of having unlimited financial resources.

Future maintenance costs should be estimated when possible, since the instrument with a higher purchase price may be more economical in the long run when maintenance is considered. The major concern, of course, is to make sure that the instrument selected meets the requirements of the system in which it is to operate.

12.5.   REFERENCES

Holmer, N.-G., 1974, "Isolation Amplifier Energized by Ultrasound," *Trans. IEEE, BME*-**21**(4), 329–333.

Holsinger, W. P., and K. M. Kempner, 1971, "Patient Electrode Isolation Adapter," *Trans. IEEE, BME*-**18**(6), 428–430.

National Fire Protection Association, 1977, *National Electrical Code*, Boston, Mass.

Roth, H. H., E. S. Teltscher, and I. M. Kane, 1975, *Electrical Safety in Health Care Facilities*, Academic Press, New York.

Appendix

A.1. OVERALL SYSTEM RESPONSE

We now present an example that indicates how analysis of overall system response can be performed. The experimental system is illustrated in Fig. A.1. It consists of a transducer driving an amplifier, which in turn drives a servomotor-operated chart recorder.

A direct approach is the use of Laplace-transform gain functions for each of the three system modules. The overall system gain function is given by

$$G(s) = G_1(s)G_2(s)G_3(s)$$

where the units of the overall gain are degrees of pen deflection per unit input applied to the transducer. Dimensionally $G_1(s)$, the transducer transfer function, is volts of output per unit input. $G_2(s)$, which represents the amplifier, is effectively dimensionless (volts per volt). $G_3(s)$, representing the servomotor and pen, is degrees per volt. Thus the overall dimension balance is

$$\deg X^{-1} = (VX^{-1})(VV^{-1})(\deg V^{-1})$$

in which X represents the units of the applied excitation.

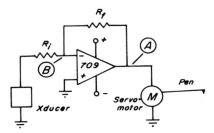

FIG. A.1   Basic transducer–amplifier–recorder instrumentation system.

We now need the signal transfer functions for the three system components. We will start with the output end and work toward the input since we have not yet specified the transducer.

The pen–motor system concept is used by the MFE Corp. (Salem, N.H.) in many of their recording instruments. This provides a much more portable (smaller size and weight) instrument than is possible with d'Arsonval-type chart recorders.

For a general servomotor driving an inertial load, such as a pen, the transfer function is

$$G_3'(s) = \frac{K_m}{s(1 + sT_3)}$$

and for our example we choose

$$G_3'(s) = \frac{100}{s(1 + s0.05)}$$

A servomotor with this type of transfer function must be used in a closed-loop feedback system. Otherwise, given a step-function input, it would rotate continuously rather than position through a fixed number of degrees. The basic feedback system is shown in Fig. A.2; K_a is a power amplifier. We will assume a wideband amplifier so that the frequency response of the system is limited by the servomotor and not by the amplifier. If the amplifier has a gain of 10, the closed loop gain (transfer function) is thus:

$$G_3(s) = \frac{G_3'(s)}{1 + \beta G_3'(s)} = \frac{K_a K_m}{s^2 T + s + \beta K_a K_m}$$

$$= \frac{1000}{0.05 s^2 + s + 100} = \frac{20{,}000}{(s + 10)^2 + 1900}$$

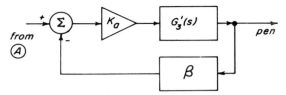

FIG. A.2   Feedback system for servomotor.

where β has been selected arbitrarily to be 0.1 V/degree (of pen deflection = motor shaft rotation). The denominator of the gain function has been expressed in the form $(s + \alpha)^2 + \gamma^2$ for subsequent convenience in finding the inverse Laplace transform.

The realization of this portion of the instrumentation system is shown in Fig. A.3. The power amplifier might be either discrete or one of the op-amp power amplifiers currently available. For this reason it is not shown in detail on the figure. The position pickoff (β box) may be either a precision single-turn potentiometer, mounted concentrically with the motor shaft, or a variable reluctance device.

The performance of this portion of the system may be determined by applying a voltage step function input at point A (Fig. A.3). Let $v_i(t) = u(t)$ V, where

$$
\begin{aligned}
u(t) &= 0 \qquad t < 0 \\
&= 1 \qquad t > 0
\end{aligned}
$$

Then $V_i(s) = 1/s$.
The position response $\Theta(s)$ is then $G_3(s)/s$ or

$$
\Theta(s) = \frac{20{,}000}{s[(s + 10)^2 + 1900]}
$$

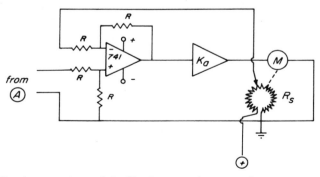

FIG. A.3   Implementation of feedback control system for servomotor. R_s is a 360° potentiometer mounted on the servomotor shaft. It provides a voltage that is proportional to the angular displacement of the motor shaft.

with the inverse transform (real time response)

$$\theta(t) = 10 - 10.26\, e^{-10t} \sin (43.59t + \phi) \text{ degrees}$$
$$\phi = 1.345 \text{ rad} = 77.08°$$

The response starts from zero and within 0.1 s has reached 63% of the final value of 10°. If we assume a pen armature length of 10 cm (a practical value), the linear deflection of the pen is $\sim 10 \tan 10° = 1.76$ cm.

We will assume that the op-amp is direct-coupled from the transducer to the servomotor–pen system. Under this condition (no interstage coupling network), an op-amp voltage amplifier has the general transfer function

$$G_2(s) = \frac{VG}{1 + sT_2}$$

in which VG is the voltage gain (1000 VV^{-1}), and $T_2 = 1.43 \times 10^{-6}$ s. Thus

$$G_2(s) = \frac{1000}{1 + s1.43 \times 10^{-6}}$$

$$= \frac{6.993 \times 10^8}{s + 6.993 \times 10^5}$$

The voltage gain of 1000 has been assumed arbitrarily (many transducers have electrical outputs in the millivolt range); T_2 for a voltage gain of 1000 was obtained from the characteristic curves for a type 709 instrumentation op-amp.

Let us now look at the system response if a voltage step function of 1 mV is applied to the input of the op-amp at point B (Fig. A.1.).

$$\Theta(s) = G_2(s)G_3(s)(10^{-3}/s)$$

$$= \frac{(2 \times 10^4)(10^{-3})(6.993 \times 10^8)}{s(s + 6.993 \times 10^5)[(s + 10)^2 + 1900]}$$

$$= \frac{1.3986 \times 10^{10}}{s(s + 6.993 \times 10^5)[(s + 10)^2 + 1900]}$$

$$\theta(t) = 10 - 4.09 \times 10^{-8}\, e^{-6.993 \times 10^5 t}$$
$$+ 10.26\, e^{-10t} \sin (43.59t - \phi) \text{ degrees}$$
$$\phi = 1.345 \text{ rad} = 77.08°$$

Thus there is some slight "ringing" of the response, but the transient oscillations are highly damped and die out very quickly. At time $t = 0.1$ s, $\theta(t)$ is $9.52°$.

The transfer functions for transducers vary substantially with transducer design. Many transducers do not produce a steady-state electrical output in response to a step function or constant input. Thus they perform only in the dynamic mode rather than in the static mode. Response, then, is proportional to the slope of the excitation and not to final value. This is true of many piezoelectric and magnetic devices. We will examine this situation subsequently.

Let us first consider a simple, linear variable-resistance transducer used for determining displacement. Such a device is shown in Fig. A.4. The transfer function is developed as follows:

For zero displacement let $R = 0$ when $l = 0$ and the output voltage $V_0 = 0$. For maximum displacement l_m, $R = R_0$, and $V_{out} = V_s$, the supply voltage.

Thus $R(l) = R_0 l / l_m$

$$V_0 = (R/R_0)V_s = (R(l)/R_0)V_s$$

which gives

$$V_0 = \frac{R_0 l}{R_0 l_m} V_s = (l/l_m)V_s$$

If the displacement is a function of time, then

$$v_0(t) = (V_s/l_m)l(t)$$

The transducer transfer function is

$$G_1(s) = V_s/l_m \qquad \mathrm{V_m^{-1}}$$

Let us assume a displacement of the form

$$\begin{aligned} l(t) &= 25t & 0 \leqslant t \leqslant 0.1 \text{ s} \\ &= 2.5 \text{ mm} & t > 0.1 \text{ s} \end{aligned}$$

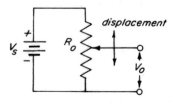

FIG. A.4   Resistive linear-displacement transducer.

that is, a linear displacement in 0.1 s to a final value of 2.5 mm. The associated transducer output voltage is then

$$v_0(t) = 25(V_s/l_m)t \qquad 0 \leqslant t \leqslant 0.1 \text{ s}$$
$$= 2.5(V_s/l_m) \qquad t > 0.1 \text{ s}$$

Let us assume $l_m = 5$ mm and $V_s = 1$ mV. The Laplace transform of $v_0(t)$ is thus

$$V_0(s) = \frac{0.005}{s^2} (1 - e^{-0.1s})$$

If we apply this voltage to the amplifier–recorder system, the output is given by

$$\Theta(s) = \frac{0.005(1 - e^{-0.1s})(1.3986 \times 10^{13})}{s^2(s + 6.993 \times 10^5)[(s + 10)^2 + 1900]}$$

$$= \frac{6.993 \times 10^{10}(1 - e^{-0.1s})}{s^2(s + 6.993 \times 10^5)[(s + 10)^2 + 1900]}$$

The pen response is

$$\theta(t) = 50(t - 0.01) + 2.92 \times 10^{-13} e^{-6.993 \times 10^5 t}$$
$$+ 1.147 e^{-10t} \sin (43.59t - \phi) \qquad 0 \leqslant t \leqslant 0.1 \text{ s}$$

Because of the numbers involved, the effective response from $t = 0$ to $t = 0.1$ s is

$$\theta(t) \sim 50(t - 0.01) + 1.147 e^{-10t} \sin (43.59t - \phi)$$
$$\phi = 154.16° = 2.69 \text{ rad}$$

when $t = 0$, $\theta = 0$; when $t = 0.1$ s, $\theta = 4.92$ degrees.
    For the range $t > 0.1$ s,

$$\theta(t) \sim 50(t - 0.01) + 1.147 e^{-10t} \sin (43.59t - \phi)$$
$$- 50(t_1 - 0.01) - 1.147 e^{-10t_1} \sin (43.59t_1 - \phi) \qquad t > 0.1 \text{ s}$$

where $t_1 = (t - 0.1)$; when

$$t = 0.1 \text{ s}, \qquad \theta = 4.92°$$
$$t = 0.2 \text{ s}, \qquad \theta = 4.54°$$
$$t = 0.3 \text{ s}, \qquad \theta = 4.99°$$
$$t = 1.0 \text{ s}, \qquad \theta = 5.0°$$

Thus the input signal (displacement) reaches its final value in 0.1 s, but the pen does not reach its final value until slightly more than 0.5 s. This is primarily a result of the 0.1 s time constant in the servomotor. Although the $\sin (43.59t - \phi)$ term affects initial response, strong

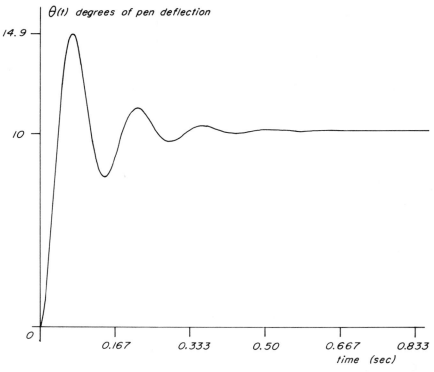

FIG. A.5 Computer plot of response of pen–servomotor system to step function input.

oscillations are not observed because of the strong damping (e^{-10t}). The several time response functions discussed above are shown in Figs. A.5–A.7. For a 10 cm pen armature, the linear deflection of the pen relative to the chart paper would be approximately $10 \tan 5° = 0.87$ cm.

Although the displacement transducer described above had a simple transfer function, with many transducers this is often not the case. Piezoelectric crystals, used in many transducer devices, have rather complex transfer functions. For a simple rectangular piezoelectric crystal (bimorph), driven mechanically to produce an electrical output, the general transfer function is

$$G_1(s) = \frac{SR_L}{S^2R_L + (R_m + sM + K/s)(1 + sT)}$$

[see Neubert (1975) for example; citation in section 4.11].

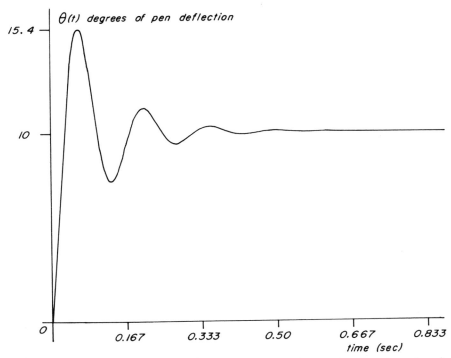

FIG. A.6 Computer plot of response of instrumentation system (less transducer) to step function input.

Some magnetic devices have similar transfer functions. The constants are defined so that K is the crystal stiffness (Nm^{-1}), M is the crystal mass (kg), R_m is the crystal damping (force damping) ($kg\ s^{-1}$), R_L is the electrical impedance into which crystal drives (Ω), S is the sensitivity factor (defined below) (Cm^{-1}), T is the time constant (defined below) (s), $S = C/d = \epsilon_0\epsilon_r A/ld$, $T = R_L C$, and $C = \epsilon_0\epsilon_r A/l$. Here, A is the cross-sectional area of crystal (m^2), C is the capacitance of crystal (F), d is the piezoelectric coefficient (CN^{-1}), ϵ_0 is the permittivity of free space $= 8.85 \times 10^{-12}$ (Fm^{-1}), ϵ_r is the crystal dielectric constant along the driven axis, and l is the crystal thickness (m). Additional relations are: $K = AE/l$, $M = \rho Al$, $R_m \sim A(\rho E)^{1/2}$, in which E is Young's modulus for the crystal (Nm^{-2}), and ρ is the density of the crystal ($kg\ m^{-3}$).

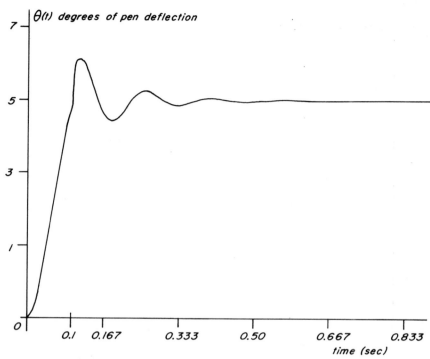

FIG. A.7 Computer plot of response of overall system to modified ramp-step function input. Note: the form of the response is extremely sensitive to computation precision of the sine function argument.

Let us assume that the transducer crystal is PZT–5A (lead–zirconate–titanate) with properties as follows:

$$A = 10 \text{ mm}^2 = 10^{-5} \text{ m}^2 \qquad E = 5.3 \times 10^{10} \text{ Nm}^{-2}$$
$$d = 3.74 \times 10^{-10} \text{ CN}^{-1} \qquad l = 1 \text{ mm} = 10^{-3} \text{m}$$
$$\epsilon_0 = 8.85 \times 10^{-12} \text{ Fm}^{-1} \qquad \rho = 7.8 \times 10^3 \text{ kg m}^{-3}$$
$$\epsilon_r = 1700 \qquad R_L = 10^4 \; \Omega$$

Thus:

$$C = 1.5 \times 10^{-10} \text{ F} \qquad R_m = 203 \text{ kg s}^{-1}$$
$$K = 5.3 \times 10^8 \text{ Nm}^{-1} \qquad S = 0.4 \text{ Cm}^{-1}$$
$$M = 7.8 \times 10^{-5} \text{ kg} \qquad T = 1.5 \times 10^{-6} \text{ s}$$

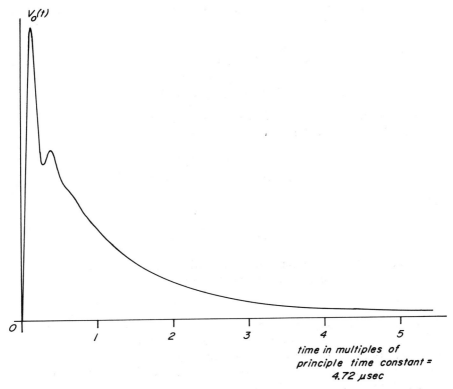

FIG. A.8 Computer plot of piezoelectric element response to an applied force step function.

The transducer transfer function can be written in the form:

$$G_1(s) = \frac{sSR_L/MT}{s^3 + s^2\left(\dfrac{1}{T} + \dfrac{R_m}{M}\right) + s\left(\dfrac{R_m + S^2R_L + KT}{MT}\right) + K/MT}$$

which becomes upon substitution of the element values:

$$G_1(s) = \frac{3.42 \times 10^{13}s}{s^3 + 3.269 \times 10^6 s^2 + 2.22 \times 10^{13}s + 4.53 \times 10^{18}}$$

This transfer function, after some manipulation may be expressed in the standard form:

$$G_1(s) = \frac{a_0 s}{(s + \gamma)[(s + \alpha)^2 + \beta^2]}$$

$$= \frac{3.42 \times 10^{13} s}{(s + 2.12 \times 10^5)[(s + 1.53 \times 10^6)^2 + 1.92 \times 10^{13}]}$$

If we now apply a step function force excitation $F(s) = F_0/s$ to the crystal, the output voltage is

$$V_0(s) = \frac{3.42 \times 10^{13} F_0}{(s + 2.12 \times 10^5)[(s + 1.53 \times 10^6)^2 + 1.92 \times 10^{13}]}$$

The corresponding output voltage is

$$v_0(t) = 3.42 F_0 \{0.478\, e^{-2.12 \times 10^5 t} + 0.5\, e^{-1.53 \times 10^6 t} \sin(4.38 \times 10^6 t + \phi)\}$$
$$\phi = -73.25° = -1.278 \text{ rad}$$

This is a very fast exponentially damped transient that rapidly decays to zero. The associated time constants are:

$$(2.12 \times 10^5)^{-1} = 4.72\ \mu\text{s} \qquad (1.53 \times 10^6)^{-1} = 0.654\ \mu\text{s}$$

Even though the applied force has a steady state value F_0, the crystal response has a steady state value of zero. The response is shown in Fig. A.8. Because of the long time constant of the servomotor, the pen would not respond to such a transient input. A very slight momentary "noise" displacement of the pen might occur.

   This last example illustrates the complexity of the transfer function characteristics of some transducer devices. Manufacturers frequently do not provide sufficient data to determine $G_1(s)$. Thus the dynamic characteristics must be determined experimentally. The discussion in this Appendix has illustrated an analytical approach, but in practice, system evaluation is usually done experimentally.

Problems

P.1. SECTIONS 1–2

1. If a mechanical transducer has the following characteristics: dynamic mass = 1.0 g, compliance = 2×10^{-5} cm/dyn, and damping factor = 0.01. What is the viscous damping in this system?

2. A transducer has the following characteristics: dynamic mass = 1.0 g, and undamped natural frequency = 100 Hz. Neglecting inertial effects, a force of 1 dyn applied to this transducer causes the dynamic mass to move with a speed of 2 cm/s. Using this information, find: (a) the compliance of the transducer; (b) its mechanical resistance; (c) the damping factor; (d) the damped frequency; and (e) the transducer time constant.

3. Assuming that the simple Nernst equation (as shown below) is obeyed,

$$\text{Emf} = (RT/ZF)\ln [X] \text{ V}$$

what potentials will a calcium ion-specific electrode produce relative

319

to a hydrogen reference electrode for the following calcium ion concentrations?

$$Ca^{2+} = 10^{-4} \qquad Ca^{2+} = 10^{-3}$$

What is the slope of the ion-specific electrode response in mV/decade? Note: $R = 8.314$ J/°-mol. Assume $T = 300$ K.

4. A monovalent cationic-selective electrode is to be used to determine potassium concentration in an electrolyte. The suspected $[K^+]$ is 5×10^{-5}. To what value must the solution be buffered? What other precautions must be observed when making this measurement?

5. A new pH meter has been installed in the clinical laboratory of a hospital. The manufacturer supplied with it a wide-range pH glass electrode and an Ag–AgCl reference electrode. The instrument is used to determine pH in a number of electrolytes including blood serum. Initially the instrument worked well and exhibited good stability and accuracy when standardized against standard pH buffer solutions. After a week of continuous operation, the instrument readings became erratic and the unit could not be standardized. Explain the probable cause for this condition and how it can be corrected.

P.2.   SECTION 3

6. The input to an oscilloscope is 1.0 MΩ to ground shunted by 10 pF. A transducer is connected to the oscilloscope by a 10-ft cable with a capacitance of 10 pF/ft. What resistance value should be placed in parallel with the oscilloscope input circuit to reduce the time constant to 1 μs?

7. Given that the transducer of Problem 6 must be terminated by a resistance of 27 kΩ or greater, design an appropriate circuit to satisfy this requirement and the time constant criterion.

8. Using op-amps, design an amplifier and filter system for amplifying an electrocardiograph signal. Specifications for the system are: input signal, 0.5–1.5 mV at approximately 1 Hz; output signal, 0.5–1.0 V; frequency response, DC to 40 Hz; 40 Hz is the -3 dB point. Frequency rolloff should be 40 dB/decade; input impedance, > 20 MΩ; and input mode, differential.

9. In an amplifier system such as defined in Problem 8, discuss the effect of a 15 mV DC offset signal applied at the input terminals. Assume that the op-amps operate from a ± 12 V bias supply. How would you eliminate the effect of this input offset?

10. Assume a thermistor with a characteristic:

$$R(T) = R_0\, e^{\alpha T}$$

where T is the temperature in degrees Celsius, α is the thermistor constant in reciprocal degrees; and R_0 is the resistance of thermistor at temperature $T = T_0$. If $T_0 = 0°C$, design an amplifier/sensor circuit that will give a voltage output *linearly* proportional to temperature change over the range 0–30°C. Commercially available amplifier chips may be used, but all circuit interconnections and additional circuit elements must be shown clearly. Additional data: $R_0 = 1000\ \Omega$; $\alpha = 0.0536/°C$. When $T = 0°$, $V_{\text{out}} = 0$ V; when $T = 30°$, $V_{\text{out}} = 9.0$ V.

11. Given the LED current–voltage characteristic shown in Fig. P.1, design an op-amp LED driver circuit according to the following specifications: LED static forward bias current = 4 mA; input

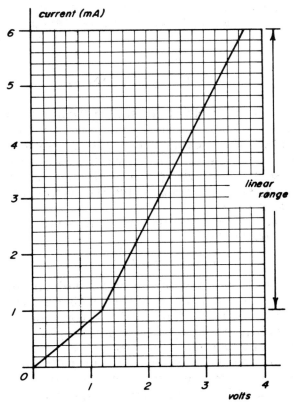

FIG. P1 LED characteristic for Problem *11*.

impedance seen by signal source driving the op-amp must equal 1.5 kΩ; amplifier gain must be adjustable from 1 to 20; op-amp bias supply (type 741) is ± 15 V DC; op-amp output should be current limited such that the LED always remains in the *linear* forward range.

12.   A constant-temperature water bath consists of a thermally insulated chamber containing a volume of water. A pump circulates the water in the chamber to maintain proper mixing. The temperature of the water is maintained by an immersible heating element that operates from 110 V AC rms. Design an electronic sensing and control system to maintain the water bath at constant temperature. The controlled temperature should be "setable" and it is not permitted to use any bi-metal elements in the system.

13.   Develop the mechanical and electronic design for a system to collect a measured volume of urine (by catheter) from a critically ill patient. When the prescribed volume has been collected, a light should go on in a control panel visible to the nurse in charge of the Intensive Care Unit. The collection assembly must be designed so that it is immune to being kicked and tipped over to avoid spilling the contents.

P.3.   SECTION 5

14.   What is the depth of penetration δ into water for a 50 MHz propagating wave if the conductivity of the water is 10^{-4} S?

15.   At 200 MHz, the conductivity of skin is ~ 0.9 Siemens and the dielectric constant is ~ 55. If a transmitting antenna coil is located subcutaneously with a receiving coil on the skin surface, what is the maximum permissible thickness of the skin (approximate value) for 200 MHz transmission?

16.   Assume a telemetry transmitter in which the oscillator frequency is determined by

$$f_0 = [2\pi \sqrt{L(C + C_v)}]^{-1}$$

where C_v represents the capacitance of a transducer and C is a fixed capacitance in the oscillator in parallel with C_v. Let $L = 0.25\,\mu H$, $C = 10\,pF$, $C_v = 0.5(1 + x^2)\,pF$, and x = conversion factor relating transducer excitation to capacitance change. Graph oscillator frequency as a function of excitation for the range $0 \leqslant x \leqslant 2$.

17.   If the telemetry receiver used with the transmitter described in Problem 16 uses a linear discriminator in its detector, what problems will be encountered in relating detector output voltage to transducer excitation?

P.4.   SECTION 6

18.   A research physician wishes to purchase a preamplifier to use in neuroelectric recording. The amplifier will be the interface between metal recording electrodes and one input channel of a Tektronix Type 3A75 plug-in amplifier module. A Tektronix Type 564 storage oscilloscope mainframe will be used. The investigator has narrowed the choice of preamplifier unit down to two devices for which specifications are shown below. Indicate which unit you would recommend that he purchase and state the basis for your recommendation.

Unit #1; cost $485.

Input: Audio phono connectors or BNC jacks; AC or DC coupling; single-ended or differential mode; two channels. Input protected by 1 A silicon diodes that clamp inputs to 1.5 V maximum and which may be removed by a front panel switch when common mode voltages > 0.4 V exist. The input impedance (resistance to ground shunted by capacitance) is given in the table below. A positive 50 μV ($\pm 2\%$) calibration voltage is provided by a pushbutton front panel switch. Maximum input is as follows: common mode, 4V (peak) positive or negative without protection diodes; differential model, 15 mV peak, positive or negative; single-ended, 15 mV peak, positive or negative; DC differential input imbalance, 5 mV (can be adjusted to zero at the output by DC zero-adjust control); common mode rejection, CMRR = 20 log (differential gain/ common mode gain) is 80 dB (10,000:1) minimum from 1 Hz to 5 kHz and 60 dB minimum (1000:1) from DC to 1 Hz.

Input impedance mode	Input protection diodes in	Input protection diodes out
AC coupled, differential	2M/120 pF	2M /50 pF
AC coupled, single-ended	1M/120 pF	1M/ 50 pF
DC coupled, differential	100M/120 pF	100M/ 50 pF
DC coupled, single-ended	100M/120 pF	100M/ 50 pF
AC coupled, common mode	0.5M/240 pF	0.5M/100 pF
DC coupled, common mode	500M/250 pF	500M/100 pF

Output: Single-ended with audio phono or BNC receptacles. Output may be shorted to the + or − power supply or to ground; short circuit current is limited to 50 mA maximum. Output voltage is 3 V peak and clipped at 6 V to protect against overvoltage damage. Maximum rated output current is 10 mA. The DC level of the output can be adjusted by 5 mV as referred to input. The output impedance is 100 Ω from DC to 5 kHz.

Frequency Response: The high and low frequency response is selected by 12-position front-panel switches. The settings are: LF (Hz): DC, 0.1, 0.5, 1, 3, 5, 10, 30, 50, 100, 300, 500; HF (Hz): 1, 3, 5, 10, 30, 100, 300, 500, 1000, 3000, 5000. A 60 Hz notch filter is built in, and may be inserted by a front panel swtich. $Q_{min} = 2$. At 47 and 77 Hz, the attenuation is 30% or less. The rejection ratio is 40dB (100:1). The filter is effective for up to 2 V peak-to-peak 60 Hz signals.

Gain: The gain (2% accuracy) is selected by a 12-position front-panel switch and a 10X or 1X switch. Gain values are: 200, 300, 400, 500, 800, 1k, 2k, 3k, 4k, 5k, 8k, 10k, 20k, 30k, 40k, 50k, 80k, and 100k. An adjustable gain control permits continuous gain adjustment from 200 to 100k.

Noise: The noise referred to the shorted input with a 5 kHz bandwidth is 8 μV peak-to-peak = 1.5 μV rms.

Stability: Short term, 8 μV (max) per 8 h with DC coupling. Long term, 20 μV per week with DC coupling. Output voltage drift, 400 μV (max) per °C with AC coupling; 1.0 mV/8 h; 10.0 mV/week. The maximum baseline variation vs line variation with DC coupling is 0.5 μV referred to the input for line changes in the range 105–130 V AC rms. The maximum DC temperature drift is 3 μV/°C with DC coupling.

Power: 105–130 V AC rms, 50/60 Hz, 4 W (max). The unit must be operated from the internal nickel–cadmium batteries in order to meet specifications. The batteries automatically charge when the unit is plugged into a power line source. Fastest charging of the batteries occurs when the power switch is in the OFF position. Charging time for fully discharged batteries is 17 h (approx).

Limits: Maximum input voltage ± 8 V. Operating temperature is 0 to +40°C; storage temperature is −20 to +60°C. Overload recovery to 1% of steady-state output is 20 s; output is saturated for a differential input signal of 80 mV.

Mechanical: Dimensions, 4-3/4″ W × 3-3/4″ D × 3-1/4″ H. Weight, 2.0 pounds = 0.8 kg.

Unit #2, cost $1075.

Input: BNC jacks; single-ended or differential mode; two channels. Input impedance (AC), through 0.1 μF shunted to ground by 100 MΩ in parallel with 15 pF; DC, 1 GΩ shunted by 15 pF. Maximum input signals: 1 V rms common mode; ± 500 mV (gains of 10–100) differential mode; ± 50 mV (gains 200–10k) differential mode. The minimum common mode signal rejection with frequency for gains greater than and less than 200, respectively, is: DC: 100 Hz, 120 and 100 dB; 1 kHz, 100 and 80 dB; 10 kHz, 80 and 60 dB; 100 kHz, 60 and 40 dB.

Output: BNC connector. 10 V peak-to-peak into 600 Ω output impedance. Distortion less than 0.01%.

Frequency response: Switch-selectable rolloff frequencies. Switch settings indicate 3 dB point of 6 dB/octave rolloff curve. LF (Hz): DC and 0.03 to 1 k in 1–3–10 sequence. HF (Hz): 3–300 k in 1–3–10 sequence.

Gain: 10–10,000 in calibrated switch settings; vernier with range expansion to 2.5X.

Noise: Noise figure = 0.3 dB at 10 Hz with 2 MΩ source impedance (max); at 1 kHz with a 1 MΩ source impedance, NF = 0.2 dB (max).

Stability: DC Coupling – 10 μV/°C or less than 10 μV/24 h (max) at constant ambient temperature as referred to input. Less than 1 mV/°C or less than 1 mV/24 h at constant ambient temperature as referred to output.

Power: Rechargeable nickel–cadmium batteries provide approximately 20 h operation between charges. Batteries charge automatically when instrument is plugged into AC receptacle. Line operation at 105–125 V AC rms or 210–250 V AC rms, 50/60 Hz, 5 W.

Mechanical: Dimensions – 8.6″ W × 4.1″ H × 11.3″ D. Weight, 4 pounds (= 1.8 kg).

Limits: Overload immediately reset by front panel switch.

19.   A patient in a cardiac care unit has been electrocuted. The following information leading to this event has been developed: The patient, a 55-yr-old male, was connected to an ECG monitoring unit, which was AC-line-operated and plugged into a wall outlet on the west side of the room. He was lying on a bed with an electric-motor-operated raising and lowering mechanism. The motor was connected to an AC outlet located on the north side of the room. The hospital wiring system was installed in 1930 when the building was constructed and had not been modified to current NFPA codes. Apparently the patient's hand slipped from the mattress and contacted a part of the metal bedframe. Comment on the probable mechanism leading to the patient's death.

20. A surgical patient under general anesthesia sustained third degree burns of the left ankle. Abdominal surgery was being performed. The patient instituted a law suit against the hospital and surgical staff in attendance at the time the surgery was performed. A consultant retained by the hospital to investigate this matter developed the following information: the metal surgical table was grounded; an electrosurgery instrument operating at 350 kHz was used during the surgery; the surgical table was equipped along the sides with metal restraining rails. Based upon this information, describe and diagram the probable mechanism leading to the patient's burn.

Index